T0231304

Flexible Manufacturing Systems in Practice

MANUFACTURING ENGINEERING AND MATERIALS PROCESSING

A Series of Reference Books and Textbooks

SERIES EDITORS

Geoffrey Boothroyd

*Chairman, Department of Industrial
and Manufacturing Engineering
University of Rhode Island
Kingston, Rhode Island*

George E. Dieter

*Dean, College of Engineering
University of Maryland
College Park, Maryland*

OTHER VOLUMES IN PREPARATION

Flexible Manufacturing Systems in Practice

Applications, Design, and Simulation

Joseph Talavage

Purdue University
West Lafayette, Indiana

Roger G. Hannam

University of Manchester Institute of Science and Technology
Manchester, England

CRC Press
Taylor & Francis Group
Boca Raton London New York

CRC Press is an imprint of the
Taylor & Francis Group, an **informa** business

Library of Congress Cataloging-in-Publication Data

Talavage, Joseph
    Flexible manufacturing systems in practice.

    (Manufacturing engineering and materials processing ;
26)
    Includes index.
    1. Flexible manufacturing systems. 2. Computer
integrated manufacturing systems. I. Hannam,
Roger.   II. Title.   III. Series.
TS155.6.H365 1988   670.42   87-30313
ISBN 0-8247-7718-2

MARCEL DEKKER, INC.
270 Madison Avenue, New York, New York 10016

Preface

This book has been written for all those interested in flexible manufacturing systems (FMS) and other forms of computerized manufacturing systems (CMS). It is thus aimed at practicing manufacturing engineers, at industrial and production engineers, at managers of these functions, and at those in programs at universities, colleges and polytechnic schools leading to first or higher degrees in these disciplines.

The book has also been written for a much larger industrial readership. The introduction of an FMS in a company is much more than the introduction of a new advanced technology system. It may represent a change in direction and approach that should be reflected right through to a company's business objectives. Ideally, implementation of an FMS should result from meeting a company's business objectives. However, a company can frame its business objectives to exploit new technology only if those at the top understand the technology and communicate those objectives so that they are understood by all employees. Thus this book should be of interest to all engaged in manufacturing.

The book is being published at a time when interest in flexible and other types of computer integrated manufacturing systems is increasing and this interest is being translated into orders for systems. The two main reasons for this are that the benefits of flexible manufacturing are being increasingly appreciated and that an FMS offers a means of implementing parts of two other modern approaches to manufacturing: just-in-time (JIT) and computer integrated manufacturing (CIM). FMS and these newer approaches have implications for those involved in manufacturing industry, and it is important that workers in all disciplines and functions understand them. In the past, few manufacturing engineers have had to be involved in the details of the design of manufacturing systems. Either the design has been subcontracted to a specialist contractor (particularly where it has involved a larger integrated system) or the manufacturing system has just evolved through purchases of individual machines. It will be shown that flexible manufacturing systems are different from other forms of manufacturing systems in a number of ways. A prime difference is that they need to be designed by a collaborative team which involves both supplier and vendor personnel. Thus many engineers now need to know what is involved in the design of an FMS.

Designing a flexible manufacturing system has many similarities with designing other products. The product design process goes through a series of stages from a tentative specification, through a layout stage supported by order of magnitude calculations, on to a detailed stage, where the design is more thoroughly analyzed and checked and optimized if possible. Those involved in the design must appreciate both the details and limitations of the hardware elements of the product, and the approaches, methodologies, and approximations of the mathematical tools used to analyze and improve the design. So it is with an FMS. It is this important combination of FMS hardware and the mathematical tools used in the design that is the subject of this book.

The book starts by putting flexible manufacturing systems into their wider manufacturing system context and by explaining the origins of FMS. Some of the pioneering systems are then reviewed and the benefits of FMS are discussed. The detailed elements and the technology of systems are presented by using examples from many systems. System hardware, software, control, management, and current developments that may affect how systems develop in the future are all reviewed.

The FMS is not just another piece of capital equipment that can be bought off the shelf. It requires the customer and vendor to collaborate closely in many aspects of system design and specification. It is here that computer modeling and simulation techniques are indispensable to ensure that the many possible combinations of the FMS hardware and storage components are investigated fully with regard to system performance before final specifications are fixed. Furthermore, the models can continue to be helpful when the FMS is in actual operation and the system supervisor needs help with planning and scheduling.

The range of modeling tools discussed in the book is wide since an FMS is such a complex system. However, an effort is made here to integrate these complementary tools into a single framework for FMS planning. Since the life cycle of an FMS includes both its design and operation, the models will find use in each portion of the life cycle. The most generally useful model types are those of computer simulation and queueing-network models. These modeling approaches are discussed as they apply to both the design and the operation of an FMS. The design process further requires economic justification of these rather expensive systems, and consideration is given to the new approaches that the peculiar (non-labor-intensive) perspective of FMS imposes on economic models. The operation of an FMS, by the definition of flexibility, implies complex problems associated with planning, loading, and scheduling. Two approaches are described for these extremely complex problems, one for formulating the problems to illustrate their complexity and the other for solving them in a realistic time frame.

Both authors have had considerable experience lecturing on FMS and simulation to students and industrialists in the United States and Europe. They have also worked closely with a number of companies in applying many of the analytical design methods so they are very familiar with the realities of designing and analyzing systems. The book reports on more than these experiences, however. In researching the industrial realities of flexible manufacturing, we have visited many U.S. and some European companies. Thanks are due to the engineers of these companies who gave their time to explain their systems and discuss their experiences. We thank The Whitworth Foundation for providing funds for Roger Hannam to undertake these visits. The companies that we visited and many other companies have supplied data, information, and illustrations for use in the book and we gratefully thank them for their assistance.

Joseph Talavage
Roger G. Hannam

Acknowledgments

We would like to acknowledge the following companies and organizations for their assistance: AB Bygg-och Transportekonomi (BI), Allis Chalmers Corporation, Avco Lycoming, British Aerospace PLC, British Monorail, Buckhardt and Weber (Hahn & Kolb), Caterpillar Tractor Company, Cincinnati Milacron, Comau SpA (RSLA Marketing International Ltd.), Detroit Diesel Allison, Fenner Systems Engineering, Fritz Werner (TI Rockwell Ltd.), General Electric, Giddings and Lewis, Inc., Giddings and Lewis-Fraser, Hughes Aircraft Company, Ingersoll Milling Machine Company, International Computers, International Hough, John Deere & Co., Kearney and Trecker Corporation, Kearney and Trecker Marwin Ltd., Leyland Vehicles Ltd., Lucas Groups Services Ltd., National Bureau of Standards, Normalair-Garrett Ltd., F. Pollard & Co. Ltd., Rockwell International, Sandvik Coromant, Scharmann Machine Ltd., S.I. Handling Systems Incorporated, Society of Manufacturing Engineers, Sundstrand Aviation, TI Machine Tools Ltd., Trumpf Machine Tools Ltd., Vought Aerospace Products, White Sundstrand/WCI, Yamazaki Machinery UK Ltd., and Mazak Corporation. Mr. Ted Holland, editor of *Metalworking Production*, is particularly thanked for permission to use a number of illustrations from the journal.

    The authors also thank for their assistance: the Department
of Mechanical Engineering at UMIST, Dr. Collin Noble, Professor
Rajan Suri, Dr. Cynthia Whitney, Dr. Ken Musselman, Professor
Richard Wysk, and the Purdue students in IE672 for reviewing
parts of the text, and Mrs. Patricia Murray for her dedication on
the word processor.

Contents

Flexible Manufacturing Systems in Practice

1

Introduction

1.1 A HISTORICAL PERSPECTIVE

This book deals with many aspects of the design, operation, and simulation of flexible manufacturing systems (FMS).* FMS represent the latest advance in types of integrated manufacturing system whose origins can be traced back to early forms of mass production. The developed world owes much of its standard of living to mass production techniques and to those engineers and entrepreneurs of the past who pioneered them.

Henry Ford is perhaps the best-known example of such a pioneer. By creating a system of manufacture that integrated and linked sequences of assembly operations, he was able to bring the purchase of a standard automobile within the reach of many who previously had looked upon automobiles as toys for the enthusiast or conveniences for the rich and well-to-do. The reduced cost of automobiles led to an expanded market for them,

---

*"FMS" will be used for both singular (system) and plural (systems).

which in turn led to growth in the size of the companies manufacturing them and in the size of workforces employed building
them. Those employed had to accept that the type of work that
needed doing within the automated environment was radically
different from what had been done before. It required different skills and different training and resulted in a different social environment at work.

Most would agree that automating the production of cars produced a large benefit for the general population. It also produced a social revolution which is still continuing, the effects
of which were, and are, far-reaching. Expanding markets for
one product often lead to growth for supporting industries so
that the initial change is doubly beneficial. One side effect, for
example, of the growth in the number of cars has been the need
to build more roads. Thus, a multibillion-dollar road construction industry has grown up on the back of the automobile industry. The need for new and better roads is still evident today.
The railroads, however, have suffered from the success of the
automobile.

Similar examples of the changes resulting from automating
manufacture could be recounted for many industries. The automation of manufacture has, almost without exception, improved
the general standard of living by reducing the cost of products.
It has also generated employment, but the type of employment
has been different and the change to automation has inevitably
had industrial and social side effects. The process of adjusting
to the social side effects has not always been easy for those involved in the change, but the number of jobs created has invariably exceeded the number of jobs destroyed.

Today's revolution in many forms of manufacturing has been
based on many factors, as will be described in Chapter 2, but
the main factors have been developments in electronics and computers. Computer developments have, of course, affected all aspects of life from the factory floor to the office and the home.
Offices are now full of word processors and personal computers,
and the two-car family has now, in addition, become the two-com-
puter family. If the computers in junior's toys and the washing
machine are included, the count will be larger. Silicon Valley is
now as famous as Detroit, with many thousands of workers doing
jobs that did not exist 10 years ago. In many respects, FMS is
based on a combination of ideas that exploit developments from
Silicon Valley and Detroit.

As will be seen, FMS can be considered the latest stage in a continuing process of improving the efficiency of manufacture through automation, although the ideas behind flexible manufacturing are more wide-ranging than this, as will be explained in Chapter 3. Although the early FMS have been in operation now for over 10 years, the use of such systems is still a new and unknown experience for many companies and certainly for the up-and-coming generation of engineers. The basic approach is relatively easy to understand — it has features of Henry Ford's assembly lines — and the benefits seem attractive. However, this level of understanding leads to an appreciation of (and often apprehension of) the complexities of FMS. A series of questions arise. What really is an FMS and how does it differ from other types of manufacturing system? How is a technical specification for an FMS developed? How can the system be matched to required production volumes? How can the system be operated to give optimum performance? Is FMS the same as unmanned manufacture? Why do some people have flexible manufacturing cells and others have flexible manufacturing systems? How does FMS relate to newer manufacturing strategies such as Just-in-time (JIT)? The list of potential questions requiring answers is vast, and this book has been written to provide some of the answers.

1.2 A PREVIEW OF FMS DESIGN

Designing an FMS requires the ability to answer the questions just posed and many similar ones. In particular, it requires a knowledge and appreciation both of the hardware of FMS and of the software tools which play an important part in their design and analysis. These tools are the mathematical and operations research techniques of simulation, mathematical programming, and queueing theory. Just as computers have contributed to the development of FMS, so their availability has contributed to the growth of software tools that enable the techniques mentioned to be used effectively on today's computers in analyzing possible designs for FMS.

Figure 1.1 illustrates a modern FMS. Its hardware comprises eight horizontal spindle machining centers, two coordinate measuring machines, a washing unit, and pallet load and unload stations with storage carousels all linked by automatic guided vehicles. One very important element of the hardware that is not

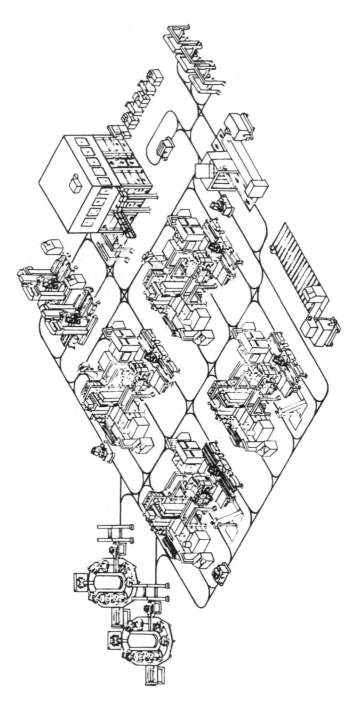

Fig. 1.1 The layout of flexible manufacturing system installed at Vought Aerospace, Dallas, Texas. (Courtesy Cincinnati Milacron.)

shown is the computer hardware, which, together with its associated software, controls the system. A more detailed description of the elements of this and other systems will be postponed until the place of flexible manufacturing within manufacturing systems has been discussed and until some of the pioneering systems that launched the elements and ideas of FMS have been presented.

In the preface, it was pointed out that the design of an FMS has similarities to other design procedures. One starts with a specification having few details and creates certain design proposals. Given some form for the design proposals, preliminary calculations can be carried out to check the feasibility of the proposals against the specification. As the layout is firmed up, more thorough analysis can start to establish more detailed proportions of a design. The design of an FMS goes through all these stages, but the procedures lead to a manufacturing system rather than product. As with designing a product, a design team must bring to bear their knowledge of both the engineering hardware and the mathematical analysis tools available to ensure the final design which will both work and be cost effective. A design team must carefully ensure their proposals are technically sound, offer economic benefits, and are financially justifiable.

An FMS is not just another piece of capital equipment that can be bought off the shelf. Such systems are complex combinations of various types of capital equipment which have to be brought together, interfaced, and made to work in unison. System complexity means that systems need to be designed by a team comprising various specialists. To match a system to a particular manufacturing company requires that the customer and vendor collaborate closely in many aspects of system design and specification. It is here that computer modeling and simulation techniques become important to ensure the modes of operation of the system are investigated fully before final specifications are fixed. Users will continue to find modeling and simulation helpful when they have a system in operation and the workpiece mix or schedules alter. For larger part mixes, continued use of simulation is essential.

FMS were originally developed for metal-cutting applications, and this is still by far the major area of their application. For this reason, the book is biased toward metal-cutting systems. However, the principles of FMS are more widely applicable, and recently the integrated approach of FMS has been applied to linking forming, shearing, and many other types of manufacturing process. The applications of simulation and modeling similarly

extend well beyond the analysis of manufacturing systems — many
types of system can be modeled.

For FMS to be completely understood, it is important that they
are explained and perceived within their larger manufacturing
systems framework. This approach is followed in this book to
ensure that the problems that FMS address are understood within a total manufacturing systems context. Traditional methods
of increasing output per employee have been through increasing
mechanization and by work simplification. Flexible manufacturing
will be shown to have its foundation in these methods but to go
one stage further by linking or integrating the simple tasks back
together. Increased automation and computer integration are
the key technologies in this, with the accent now being on increasing the output and efficiency of systems rather than of individual workers.

There are few step changes in technological development; rather it is a progression. The development of FMS was a progression. This will be illustrated by describing some of the early integrated systems that had many of the elements of today's FMS.
Once the overall concepts and form of flexible systems have been
presented, the details of many aspects of the hardware of FMS
will be discussed to give the necessary background for considering alternative system configurations and how FMS are controlled and manned.

Modeling and simulation will be shown to be vital tools at a
number of stages in the design process. The presentation of
these topics has been left to the second part of the book as it
is considered important to understand what has to be simulated
and modeled before discussing how it is carried out.

1.3 MANUFACTURING SYSTEMS

The term "manufacturing" probably embraces a greater range of
activity than any other word in the English language. It can refer to the production of nylon yarn or of textiles, to the production of pins or of steel, and to the production of the millions of
individual parts that are assembled in combinations to form the
vast diversity of products that exist today. The products may
be comparatively simple, such as a hand-held drill, or have the
complexity of a gas turbine engine or a jumbo jet on which the
engine is mounted. Manufacturing can also refer to the production of plastics and other chemicals that are produced by flow

and reaction processes and even to the processing of food and confectionery.

This text is concerned with systems for the manufacture of products that have a number of constituent parts, generally made of metal. Our interest further concentrates on that part of the total manufacturing process that converts raw materials into finished components and then products. The raw-material form of the components will generally be castings or forgings, or rolled or drawn bars or plate. Producing the finished components requires a whole series of operations to be carried out on the components. These may include any of a large range of forming and machining operations, as well as finishing operations and allied operations such as washing and inspection. Once finished, the components will generally be assembled into subassemblies before these are brought together in the final assembly of a product. Although the text will concentrate on the manufacture of products with metal parts, many of the ideas and mathematical techniques presented can be applied far more generally.

Manufacturing systems used to produce the products are usually classified under three headings on the basis of the quantity produced at one time. Thus, there are systems for mass production, batch production, and jobbing production. As will be seen later, these may be further classified and characterized by the physical layout of the machines.

Mass production is manufacture in which products are produced continuously by special-purpose machines. The machines can be specialized since they are not required to produce anything else, and considerable ingenuity and expense can be put into their design and manufacture to ensure they operate effectively. Mass production has been characterized by having "fixed automation," that is, automation that cannot be changed, and the machines are referred to as "dedicated" to their specialized task.

Although the products of mass production are plainly visible on our roads and in our homes, most products are not required in such quantities and are produced by batch production methods. This involves producing components and products in batches that may range from say 10 to 10,000 at a time. The machines used to produce the components must be more general purpose than those used for mass production and must be capable of being set up in different ways with a variety of tooling to produce a range of components for a number of different products. Specialized tooling, fixtures, and gages may well be used on the machines depending on the total component quantities required.

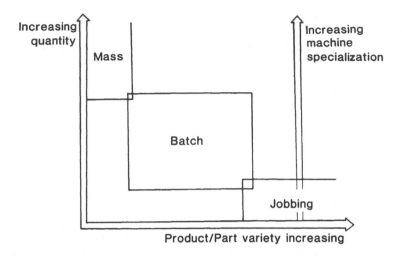

Fig. 1.2 Production type, variety, machine, and quantity relationships.

The greater the quantity, the more specialized the equipment can be because it will be financially worthwhile. Batch production is estimated to account for over 70% of all production in developed countries.

Some products and components are required only in very small quantities, and these are produced by jobbing methods. Specialized tooling, fixtures, and gages cannot usually be justified for small-quantity production, and general-purpose machines and tooling are used together with manual methods. Examples of products that are required only occasionally or in very small numbers are special machines, process plants, machine attachments, prototypes, and certain large machines.

Figure 1.2 summarizes the relationship between the types of production, the specialization of the machines, and the quantity of production. The boundaries between these three classes of manufacturing system are somewhat ill defined, and some writers might insert a high-volume class of system between the batch and mass production classes. It should be pointed out that a given type of product can be produced by methods from all three classes, depending on the size of the market. Products produced by jobbing methods are generally far more expensive than those produced by mass production methods, but they meet the needs of the more sophisticated and exclusive end of the market where customers are able to pay higher prices. As will now be explained, flexible manufacturing is not a system in the sense of the three systems of manufacture just described, it is rather a philosophy or an approach to manufacture that can be overlaid across these classes. Indeed, FMS have now been applied to mass production, to batch production, and to the production of a succession of individual piece parts.

1.4 FLEXIBLE MANUFACTURING

The adjective "flexible" only started to be applied to manufacturing systems in the late 1960s and early 1970s, when it was applied to various novel integrated manufacturing systems that were developed at the time. These were all systems for cutting machine parts, and some of their features will be described in Chapter 3. By the mid-1970s, these and other similar systems became generally known as flexible manufacturing systems, or FMS. The systems embodied a manufacturing philosophy and certain ideas that were quite revolutionary at the time — as were the systems themselves.

In the early 1970s, the philosophy of flexible manufacture and its implementation through FMS were viewed as one. However, it did not take too long for manufacturing engineers to apply the philosophy to entirely different arrangements of machines, but, for want of another word, these were also called FMS. Variations on the acronym FMS have since been introduced with the use of FMC, standing for flexible manufacturing cell, and FFS, flexible fabrication system.

It might help the reader and would certainly help the authors if flexible manufacturing could be comprehensively defined on page 1 of the first chapter of this book and a "top-down" description of it developed from the definition. Flexible manufacturing has, however, grown from a number of roots which have all come together, and a complete understanding of what flexible manufacturing is, what it sets out to achieve, and the diverse forms of its implementation can only be attained if its roots are understood. A "bottom-up" description of its development thus must be given. This will enable flexible manufacturing to be placed in its context as the manufacturing philosophy for the remaining years of the 20th century and probably for the early part of the 21st century.

To illustrate the difficulty of defining flexible manufacturing, a definition of it is now given, but the statements will not, at this stage, be explained or expanded.

A manufacturing system is flexible if it is capable of processing a number of different workpieces simultaneously and automatically, with the machines in the system being able to accept and carry out the operations on the workpieces in any sequence. For this to occur, the machines in the system must, on completion of the part currently being processed, have available the next workpiece, the tools to process the next workpiece, and the necessary instructions, so a simple changeover of workpieces, tools, and instructions can occur, normally without manual intervention.

It can be seen that this is a complex statement, and it is more a description than a definition. This reflects two realities of FMS, namely, that they are complex and are easier to describe than to define. This topic is returned to, however, in Chapter 3 once the forms of some systems have been described.

If an informed industrial archaeologist were to view any current system 150 years from now (if it was still available to be viewed), he could probably date its installation to within 5 years. This short time period would be possible because developments are currently moving forward relatively rapidly, particularly when the means of controlling systems are considered. If the archaeologist were to try to seek reasons for the systems, perhaps by reading an adjacent journal or book, he would find the principles behind a system of the 1980s would be very similar to those of the 1960s, of the 1930s, and even of the 1880s. In broad terms, the principles relate to the need to minimize costs and maximize productivity, but each has to relate to the technology and manufacturing practice and economics of the time. This book deals with the same principles but looks at means of implementing them for the remainder of the 20th century with current technology and in response to the present market and economic scenario.

The fact that manufacturing engineers have been working toward the same goals for the last 150 years is significant because some of the means of achieving our present goals have been around for a long time. Equally, some of the approaches taken today, and particularly the flexible manufacturing approach, are radically different from those taken before because we have the technology to do things differently. The development of today's systems from their earlier origins is worth reviewing to get a complete picture of how systems developed to where they are today. The roots of flexible manufacturing lie in earlier systems, and there are lessons for today and tomorrow from the approaches taken in the past. These are the topics of the next chapter.

2

The Development of Manufacturing Systems

2.1 THE COMING OF SPECIALIZATION

The story of the development of manufacturing industry is a fascinating one. It is a story of human endeavor by great engineers most of whom were also very successful in business as well as men of vision. The brief introduction to the history of manufacturing systems that is included here is very selective in the parts of the history presented. The parts chosen link through in some way to today's developments, particularly to the development of FMS and modern integrated manufacturing systems.

A manufactured article is one made in some form of factory or plant in an organized way. It contrasts with "home-made," which means that complete products are produced by one person or a very small group of people in their homes. Until the middle of the 18th century, almost all production was home-made in what became known as cottage industries. Each cottage was run by an owner—craftsman who was self-employed and who brought a number of skills to bear in the manufacture of his product. He generally bought his raw material and sold his finished products through merchants at fairs.

What has been termed the Industrial Revolution occurred between 1750 and 1850, when a number of commercial and technological factors had developed to a certain stage. Three technological factors were the development of powered machinery, the harnessing of water and then steam power to drive the machines, and the development of means of transport to handle the larger quantities of raw materials. Related factors in Britain included the existence of merchant entrepreneurs and the merchant venturing tradition, which enabled the entrepreneurs to finance their ideas.

Factories developed because the powered machines that were developed were expensive and could only be afforded by the merchants. To increase their business, the merchants initially loaned the machines to the cottage-based craftsman, who thus became subcontractors to the merchants. The merchants soon realized that they could supervise their subcontractors and their machines more effectively if they were housed together in a single building. So started the factory system and the separation of capital and labor, owner and employee.

The new machines were designed to take over tasks that had previously been done manually and to do them faster. The operator and his machine thus became far more productive. However, the machines that were developed were rarely designed to do more than one task; thus those who operated the machines became specialized operators. This approach followed that advocated by Adam Smith in his *Wealth of Nations*, published in 1776 [1]. He saw the "division of labor," that is, labor specialization, as leading to increased productivity. As he wrote:

> This great increase of the quantity of work, which in consequence of the division of labour, the same number of people are capable of performing, is owing to three different circumstances; first to the increased dexterity in every particular workman; secondly, to the saving of time which is commonly lost in passing from one species of work to another; and lastly, to the invention of a great number of machines which facilitate and abridge labour and enable one man to do the work of many.

Smith's ideas might seem unremarkable. His own name suggests an earlier specialization in his family, and since the coming of civilization, men and women have adopted specialized occupations. However, Smith's ideas involved a considerably greater degree of specialization within occupations than was common at that time.

In the emerging factories, similar types of machines were grouped together so that operators with similar skills worked alongside one another. This layout of manufacturing is now termed "functional" because machines with similar functions are grouped together. Functional layouts were initially used in textile mills, but later they were also applied in the emerging machine shops that produced the machines for other industries to use. Thus, in a machine shop, there was an area for chucking lathes, another for center lathes, an area for milling machines, another for boring machines, and so forth. The layout of machines in functional groups is still very common in batch manufacturing today.

2.2 THE BEGINNING OF MASS PRODUCTION

It is interesting to note that a 1835 definition of "a factory" by Ure [2] included the statement that a factory had "a system of productive machines." Such factories had existed by Ure's time. A noteworthy example of 1808 was a factory in England for producing pulley blocks for the British Admiralty. This reportedly produced 130,000 blocks a year using 44 machines (built by Maudslay) and manned by 10 unskilled operators. The output from the factory exceeded the output of the six dockyards and the 110 skilled workers who had previously produced the blocks manually [3]. This is one of the early examples of mass production, in which a series of dedicated machines continuously produce a single product. It is interesting to note that the development took place in response to military requirements, and this feature has characterized many manufacturing developments over the centuries.

Another example of this occurred in the United States at the end of the 18th century and involved Eli Whitney, an engineering entrepreneur who had already made his name by inventing the cotton gin [4]. At that time, the newly independent U.S. government, worried by the intentions of France and Britain, wished to have a home-based facility for manufacturing muskets. It also required that the muskets should be so made that any musket damaged in the field could be easily repaired on the spot without the need for skilled fitting by scarce gunsmiths. This requirement demanded that parts be easily interchangeable and thus made to close tolerances. The U.S. government was aware that the French had achieved the standards necessary for interchangeable manufacture and wished to emulate them.

Eli Whitney convinced the government he could supply what was required, and in 1798, he was awarded a contract to supply 10,000 muskets that would have interchangeable parts. Whitney had considerable difficulty fulfilling the contract, but he kept the government satisfied to the extent that further contracts were awarded. Initially, parts were hand-filed to precision jigs and gages, but this was slow and encouraged Whitney to develop machines (including the first milling machine) that would be faster and more accurate and so enhance the rate and quality of production. Whitney and other American inventors and entrepreneurs had to concentrate on designing machines to carry out their more critical manufacturing operations because they did not have access to a workforce of skilled craftsmen, as had developed in Europe. This reliance on machines and increasingly mechanized operation progressively gave the United States a lead over Europe in the application of machines and mechanization to manufacture.

At the beginning of the 19th century in Britain, the factory system and mass production were primarily being developed in the textile industry. Few other products warranted being mass-produced. The inventions of Kay (flying shuttle), Hargreaves (spinning jenny), Crompton (mule), and Arkwright (water frame) are well documented, and the machines they designed were built by a skilled workforce of engineering craftsmen. These men were led by practical engineers, such as Wilkinson, Maudslay, Nasmyth, and Whitworth, who not only built machines for others but also made significant contributions both to improving existing machine tools and to developing new types [5].

In the United States, the path trodden by Whitney was quickly followed by others, with the result that similar manufacturing approaches were applied to the production of certain consumer articles [4]. The process was taken further by Samuel Colt in the 1830s, allied to the manufacture of his revolver. Once again, it was a need for arms — this time for the war with Mexico — that was the spur behind a technological advance. The American lead in automated manufacture was recognized in Europe by the method of producing parts accurately on semiautomatic machines being termed the "American System." One aftermath of the American Civil War was a continued shortage of skilled labor, and this spurred the development of further automation in many types of industry and gave the United States a lead in the application of automation to production which has only recently been challenged.

Simple items can be mass-produced on single machines, whereas more complex items require a linked sequence of operations to be carried out on them. The linking of sequential operations and production by this method is generally known as flow-line production. A description that could apply to flow-line production can be found in Adam Smith's writings, but in the context of the division of labor that has previously been mentioned. He wrote, when referring to the manufacture of pins from wire [1]:

> One man draws out the wire, another straightens it, and a third cuts it, a fourth points it, a fifth grinds it at the top for receiving the head; to make the head requires two or three distinct operations; to put it on is a peculiar business, to whiten the pins is another; it is even a trade by itself to put them into the paper; and the important business of making a pin is, in this manner, divided into eighteen distinct operation. ...

This description does not mention any means of transferring the pins from man to man, which is an essential element of flow-line production and which today is primarily associated with the manufacture of automobiles. An example of flow-line production is known to have existed in 1833 in a factory run by the British Admiralty for making ships' biscuits [3]. Work was moved between workers on trays mounted on powered rollers. In the United States, a significant development in the 1860s was an overhead conveyor line used for butchering pigs. The conveyor carried the carcasses past a line of workers who each had a specific job to do. The system even included an in-process station for weighing carcasses [3].

In 1886, Karl Benz and Gottlieb Daimler launched the age of the automobile with their horseless carriages, and other engineers were soon developing similar vehicles in most European countries and in the United States. The early vehicles were almost individually built, but it did not take long for the techniques of mass production to be applied to their manufacture. The manufacturing expertise and tradition of the United States were particularly relevant to this task. Many of the techniques of interchangeable manufacture had been developed and applied in the United States, and it had some experience of using flow-lines. The years 1850–1900 had also seen considerable innovation in and development of the more specialized type of machine tool. These included universal milling machines, drilling machines, automatic lathes, grinding machines, and various types of gear-cutting machinery, as well as machines for forming and die casting, all of which were necessary for the successful manufacture of cars.

A significant milestone for integrated manufacturing occurred in 1913/14 when Henry Ford developed a flow-line on which an engine was progressively assembled in 84 stages. Ford's assembly line reduced the labor required to build an engine by two-thirds and also increased the rate and discipline of engine production. Applying similar principles to other stages of car assembly helped reduce the cost of a car dramatically and established flow-line manufacture as the most cost-effective approach to mass production [3].

Although Henry Ford is known as the father of the assembly line, he did not invent it. Adam Smith had expounded the ideas embodied in the assembly line over 130 years before Ford, and as has been explained, systems had existed previously for conveying work between workers. Henry Ford's contribution is significant because of the scale of his assembly lines and the novelty of the application. He demonstrated that the combination of the division of labor and flow-line production offered significant savings in the cost and time of manufacture. Ford's approach was further reinforced in 1917 by F. W. Taylor [6]. Among his many other contributions to scientific management, Taylor can be considered the father of time and motion study because he scientifically analyzed what a worker could do and how best he might do it. This was particularly relevant to the working of assembly lines. The division-of-labor ideas of Adam Smith were thus taken a stage further and reinforced as the means of effective manufacture.

2.3 TRANSFER MACHINES AND LINK LINES

A bridge between these developments and FMS came with the advent of the transfer machine and later of the link line. In transfer machines and link lines, the principles of flow-line manufacture are applied to machining operations. With transfer machines, high-volume production is achieved by dividing the machining of a part into a number of short operations, which are carried out sequentially at a series of specialized workstations. The workstations are arranged in a line (or a circle) and parts are simultaneously machined, the particular machining operation being dependent on how far a part has progressed along the line. At the end of each machining cycle, all parts are simultaneously transferred to the next workstation, with a finished machined part coming off the end of the line and an unmachined part being loaded at the first workstation. Station cycle times can typically range from 20 seconds to 2 minutes, which means that parts are produced at this rate. The number of stations can range from about six up to over 50, depending on the machining to be done and the production rate required.

The design of a transfer machine requires that the operations at each workstation be reasonably equally balanced in the time they take. Every extra station incurs an extra cost, so it is important that the cutting cycle at each station be as nearly equal to the cycle time between transfers as possible. Figure 2.1 illustrates a modern transfer line for machining the rear axle housings of a truck. A number of boring stations can be seen in the foreground and stations with angled drilling heads in the background.

The development of the first transfer machine is credited to Greenlee and Co. in 1908 [7]. However, this was only a two-station machine with manual transfer between stations and was used only for boring and adzing wooden railway sleepers. The first metal-cutting transfer line was not developed until some 15 years later, when the Archdale company of England built a 53-station line for the machining of Morris car cylinder blocks. This was a notable advance, but it was still not a fully automatic transfer line as it required manual transfer of parts between stations. It did not take long, however, for completely automated lines to be produced.

Transfer machines are generally used for carrying out drilling, tapping, reaming, boring, milling, and broaching operations on components that are initially produced as castings or forgings, as illustrated in Fig. 2.1. Workstations generally have only a single axis of movement because this is all that is required for such operations. If milling and broaching are required, then two axes of movement may be required or the workpiece may have to feed past a machining head. Workpieces are usually mounted in a fixture on a pallet, and they are transported through the machine and machined at each station while retained on the pallet. Simple workpieces may be just clamped to the pallet, although more complex types require a pallet and a fixture or an integral pallet-fixture. Figure 2.2 illustrates a pallet-fixture for holding a rear axle housing though not for the housing illustrated in Fig. 2.1. If a workpiece has some naturally flat surfaces, these may be premachined and adapted for location at the workstations so the use of a pallet is avoided. A discussion of pallet and fixture configurations for flexible systems is presented in Chapter 4.

Fig. 2.1 A modern transfer line for machining the rear axle housings of a truck. (Courtesy Kearney and Trecker Marwin.)

Fig. 2.2 The loading position of a transfer machine showing an empty and loaded pallet-fixture. (Courtesy Kearney and Trecker Marwin.)

A feature of transfer machine design that aids their productivity is the use of unit heads – sometimes referred to as multi-spindle heads. Three unit heads can be seen in Fig. 2.3, which shows part of a Burkhardt and Weber transfer machine. Unit heads consist of a gearbox which drives and has protruding from its front a number of fixed spindles. These may typically hold drills, reamers, taps, or boring tools. The spindles are permanently fixed and positioned to correspond to the pattern of bores or holes to be produced in a part at a particular workstation. The unit head to the right of the figure carries 29 drills, whereas the other two unit heads both have a number of boring spindles. Unit heads are very productive compared with drilling or boring

Fig. 2.3 Unit heads forming part of a Burkhardt and Weber transfer machine. (Courtesy of Burkhardt and Weber and Hahn & Kolb.)

machines, which have only a single spindle. The machining of a
number of bores or holes simultaneously not only shortens the
production time for a part, but also reduces the number of stations required in a transfer machine. The part being machined
in Fig. 2.3 is hardly visible as it is held within a box type pallet-fixture. This type of fixture tends to be used when a part
is reoriented at some stage in its machining.

Parts that require turning, cylindrical grinding, gear cutting, or similar operations are not generally machined on a transfer line because they require rotating in different orientations for these operations. Such workpieces can be mass-produced on link lines. A link line has the required machine tools (i.e., the lathes, grinders, hobbers) arranged beside and linked to a work-handling system. The work-handling system is often a conveyor. A pick-and-place unit is used to transfer the workpieces from the conveyor to the machine chuck or other form of fixture that automatically clamps the part in the machine. The machines are set up to carry out the necessary operations, and their controls are interlocked with those of the work-handling system. Parts may or may not be mounted on pallets, depending on their shape. Link line pallets only need to provide a simple means of support for the parts and to orient them for the pick-and-place unit to load into the machine. FMS are a form of link line, as will be seen when various examples are discussed and illustrated in subsequent chapters.

Thus, the flow-line principles behind transfer machines and link lines are the same. The main difference in the application of the principles is that link lines incorporate complete machine tools, whereas transfer machines have specialized machining modules. Both types of line may have nonmachining stations in them. These stations may be for washing, gaging, reorientation of the part, testing, etc. Consequently, both lines can be considered as complete processing lines rather than just machining lines.

2.3.1 Transfer Line Economics

Transfer machines are expensive items of a plant. Although each machining station only has to carry out the same operation time after time and can therefore be comparatively simple, the linking together of the many stations with the transfer mechanisms and the provision of the integrated control system, coolant system, and swarf-handling system makes for a substantial and complex piece of capital plant. Such equipment needs to be kept working, typically on three-shift operation, to earn its keep. If it is kept working, the speed of manufacture in terms of number of parts produced per hour means that the cost per part can be very low. The important factors that contribute to this are the following:

1. The workpiece transfer time is a small proportion of the cycle time at each station so the ratio of cutting time to noncutting

time is high and the time wasted as workpieces pass through
a machine is small.

2. Time lost by workpiece loading and unloading is minimized by
these "operations" being carried out in parallel with cutting,
effectively forming the first and last stations of the line.

3. The continuous nature of mass production, its predictability,
and the lack of workpiece variety requiring changed setups
mean that machine utilization of over 85% is often achieved.
This is a significant contribution to low production cost.

4. The dedicated tooling at each workstation, coupled with high
machine utilization, gives high tooling utilization.

5. The machine station design permits two (and sometimes three)
sides of a workpiece to be machined simultaneously, often
using unit heads. This concentration of cutting power helps
reduce machine cycle times as well as the number of workstations required.

6. A machine with many stations can be looked after by one man.
He may be assisted by others when changing tools, but when
the machine is producing, a single "operator" is all that is required.

7. The integrated work-handling system together with the machine design enables a series of operations to be linked, elim-
inating any need for scheduling between the operations.

These features of manufacture by transfer machine apply to
most link lines as well. When coupled with three-shift operation,
they give a lower cost per part than other means of manufacture.
Low part costs are only achieved, however, if high utilizations
are maintained over the writeoff period of the machine, which
may be 5 years. This requires confidence on the part of the pro-
ducer that the market demand for the product being manufactured
will continue and that the part is not going to be subject to engineering changes. Such changes could easily make a transfer machine prematurely obsolete. This is because a transfer machine
is a highly specialized and dedicated piece of fixed automation
which has been designed to do a set series of manufacturing operations, and only those operations. To attempt to change it is
both difficult and expensive. In fact, it may cost as much to
adapt an existing transfer machine to a modified part design as
it would cost to buy a new machine for the modified part. Although fixed automation was a characteristic of transfer machines
up to the end of the 1960s, the 1970s and 1980s have seen the development of transfer lines that can accept some product variety
(see for example Ref. 8).

One feature of the immediate post-World War II years was that the world market bought what was then available. By the mid- to late 1960s, however, consumers started to have an increasing choice of products and started to exercise it. Manufacturers began to realize that long runs of the same product could no longer be guaranteed. Products had to be remodeled more often to compete in an increasingly competitive world; so there occurred a reduction in the number of manufacturers who were able to accept automated plant that could not be reset. Manufacturers had to start to plan for a degree of product variation within one product range, perhaps offering four or five variants of the same basic design. It was this market push that led to flexible manufacturing.

2.4 BATCH PRODUCTION

Although the products of mass production are evident in every home and on the roads, mass production accounts for a small proportion of most developed countries' total manufactured output. The major part of the manufactured output, accounting for more than 70%, is produced by batch manufacturing methods [9]. As explained in Chapter 1, batch production is the method adopted when the required product volumes are not adequate to permit continuous production of one product on dedicated machines. More general-purpose machines have to be used, which can be reset to produce a range of different parts in various batch quantities. This ensures the machines can be kept busy and thus earn their keep. Batch production does not, therefore, benefit from the factors that give mass production its low costs.

The development of batch production cannot be chronicled as the development of a system in the manner possible with mass production. Stages in its early development relate more to the development of the many forms of machine tool that are evident in manufacturing industry today. Development in this century has taken place on two fronts. First, improvements have been made in the technology both of the machines (including developing completely new types of machines) and of the tooling they use. Second, improvements have been made in the organization of production. Both these developments have been designed to overcome some of the significant problems of batch production. These problems will now be discussed.

The organizational differences between producing one part continuously and producing batches of many different parts are very large and very significant. Different sets of tooling, different gages and fixtures, different workpieces, different process plans, different drawings and job order numbers — all these have to be organized and brought together at a large number of different machines throughout a machine shop. There are inevitable delays and problems. Once all the requirements have been brought together, the setup for the previous batch has to be stripped down and the new setup made with the new part and with different tooling and different mechanical and electrical settings for the machine. The first part of the new batch then has to be carefully produced and checked before the production of the new batch gets fully underway.

In batch production, one operator is generally allocated to each machine. If the operator leaves a machine for a period of time for any reason, he will stop his machine and its utilization will be adversely affected. Batch production provides many opportunities for an operator to leave his machine to sort out manufacturing difficulties. Whereas it is a recognized fact of life of mass production that machinery must be kept working, batch production does not have this tradition. Thus, machines may be operated for only one or two shifts a day, and they may be idle every weekend and during holiday breaks.

A common means of scheduling batch production in machine shops is on the basis of an operation per week. Thus, a batch that has to visit 20 different machines for its operations will take 20 weeks to be completed. The reason for this approach is the organizational difficulty of controlling and monitoring the progress of batches through a machine shop. Even on a weekly schedule basis, it is necessary to have progress chasers or expediters to keep batches moving. The only reason for having expediters is that the organizational system does not work adequately. The expediters' wages are an extra cost of that inefficiency. But extra costs do not stop there. Expediters need to work from shortage lists which have to be produced. A priority system for late batches may have to be run alongside the scheduling system to further complicate matters, with color stickers indicating how urgent a particular batch is. To get the shortages produced, expediters have to work with section foremen, taking up their time and often requiring them to revise their existing plans, all effectively incurring further costs. And so the story could be continued.

There are other financial consequences of the organizational difficulties of batch manufacture and the "one-week-per-batch" procedure. First, for the machine, it has been found [10] that the ratio of cutting to noncutting time is very low and machine utilization at best may typically be only 30% — it may be a lot lower. Low machine utilization means that more machines and operators are required to produce a given output than would be required if utilizations were higher, with resulting increased costs. If the situation is viewed in terms of the workpieces, then parts produced by batch production methods are found to spend the majority of the time they take to be produced just waiting on the shop floor. They may be waiting for the completion of the batch, for inspection, for transport, for documentation, for the next operation, and so forth.

This interrupted form of production with parts/batches having to wait between operations contrasts sharply with the continuous nature of mass production. It results in large volumes of work-in-progress (WIP) which need financing by increased working capital. It requires space to accommodate the WIP and staff to organize it. It leads to high inventory, a slow turnover of stocks, and poor cash flow. All these factors add to the cost of production. Companies in this situation cannot respond to market changes quickly and thus may be left with outdated products in stock while being unable to get new products to the marketplace quickly. Because the batch size will generally represent the requirements for a part over a period, production in batches always generates inventory. That is, some parts need to be stored when finished because the market demands have been exceeded by the batch quantity produced. There is considerable debate on the size of batches to schedule, but batches are necessary for economic and logistical reasons.

It has already been explained that batch production involves resetting machines between batches. This may take half an hour on a simple machine. On more complicated automatic machines, such as multispindle lathes, it may take 8 hours or longer. Machine usage is costed at a given rate per hour, and setting up is a nonproductive activity that has to be costed against the workpieces produced by the setup. Thus, the larger the batch size, the smaller the setup cost contribution for each part. However, large batches incur inventory costs for storage and for financing, and these costs will be high when the inventory is in the form of finished components or products with a high added value (that is, parts that have been through expensive manufacturing operations).

The 1960s saw three significant attempts to tackle the problems of batch production (or at least some of the problems). These were the application of group technology, the use of numerically controlled machines, and the application of data-processing computers.

2.4.1 Group Technology

The philosophy of group technology (GT) was first proposed by Mitrofanov in 1938 [11], although it was not until the 1960s that it started to receive much attention outside the U.S.S.R. It is based on minimizing a part's handling time by carrying out all of its operations within a single group (or cell) of machines which are arranged adjacent to one another. To apply GT, all the parts produced by a machine shop must be sorted to create "families" of parts. A family is distinguished by having similarities in size, geometry, and methods of manufacture. The analysis required to create families of parts can be quite time consuming, and it is not uncommon to use a part coding and classification system to help in the analysis. The operations on each part family have to be related to particular machines, which thus form a cell. The number and types of cells for the whole shop have to be determined. The shop then has to be reorganized with the machines laid out in their appropriate cells. It can be appreciated that the implementation of GT can take some time.

The use of GT is designed to give a number of benefits. It eases and simplifies scheduling problems in batch manufacture because batches only need to be scheduled to one cell rather than to a number of machine groups, as in functionally laid out machine shops. This avoids the one-week-per-operation norm of many batch operations, thus leading to a quicker processing of parts and lower WIP. Expediting, if necessary, becomes straightforward, as a batch will only be in a particular cell whereas in a functionally laid out shop, if a particular batch is not being machined, it could be almost anywhere in a shop, depending on where a fork-lift truck driver has found space to put it. Similarly, parts only require to be taken to a cell as raw material and collected from it as finished components. Thus, part handling, like scheduling, is considerably reduced and simplified. Additionally, the similarity of the parts within the cell family means that fewer tools need to be changed or adjusted for setting up a new part, thus reducing setup time at the machine. All these factors simplify the task of organizing manufacture within a GT-based machine shop and speed the production of parts produced by it.

Group technology can be described as an approach to making batch production a form of flow-line production because parts,

once in the cell, progress from machine to machine and then leave
the cell. Although parts still do this in batches, the reduced setup time and the reduced through-put time both mean that it is
often practicable to reduce batch sizes, giving a more continuous
type of production. Most GT machine shops have manual part/
batch handling within a cell and rely on the proximity of the machines to make this possible. Some, however, have roller conveyors linking machines. This has the benefits of both easing
handling and making all WIP very visible.

2.4.2 Numerical Control and Machining Centers

GT is an organizational approach to overcoming the organizational
problems of managing batch production. The application of numerical control is a technological approach to tackling some of the
problems associated with batching. The two main problems tackled
are the length of time it takes to set up a machine (which in turn
means large batches and increased inventory), and the delays
generated by the one-week-per-batch mode of operation (which
cause long manufacturing lead-times).

The first numerically controlled (NC) machine tool was developed in 1952 at the Massachusetts Institute of Technology for
machining complex profiles. The reasons for its development had
nothing to do with batch production problems, but with the need
to be able to machine complex profiles consistently. However,
the potential benefits for batch production of using NC machines
were soon realized, although this did not speed their application
into industry.

NC machines offer batch manufacturers a number of significant
advantages:

1. Complex (or simple) part geometry can be produced without
    the use of expensive jigs to help position tooling. Jigs had
    often been necessary with conventionally controlled machines.
2. By capturing the part geometry through the machining in-
    structions on coded magnetic or punched-hole paper tape,
    much of the dead-time associated with setting up a machine
    is eliminated.
3. Machine repeatability and the use of identical instructions
    mean that the part geometry can be produced repetitively
    and consistently once a tape is proven. (Final part accu-
    racy is also governed by tool wear.)
4. Reduced setup time as a proportion of the total batch manu-
    facturing cycle time makes it economical to machine smaller

batches. Smaller batches reduce WIP, with the advantages
already described.

Early NC machines imitated the functions of their conventional
counterparts. Thus, there were NC drilling machines, NC milling machines, and NC boring machines. The development of the
machining center that could drill, mill, and bore was very significant for batch production because it combined operations and
enabled previously separate operations to be joined and carried
out at one setup. The dead-time associated with the previous
"extra" operations and the time lost between these operations
were thus eliminated, with great advantages as far as the total
manufacturing cycle time for a part was concerned. Every operation eliminated was potentially a week saved. This, in turn,
improved product lead times and reduced WIP.

The early machining centers were mainly vertical spindle machines and were developments of vertical drills. The horizontal
machining center (HMC) was a development of larger horizontal
boring machines. By the end of the 1960s, the design and operation of machining centers had been enhanced by the provision
of automatic tool changers (ATC), changeable worktables, and
indexing worktables. An HMC with all these features is illustrated in Fig. 2.4. Automatic tool changers were applied to both
horizontal and vertical machining centers, and they significantly
reduced in-cycle dead-time compared with the early designs of
machining centers, which required an operator to change tools
manually at a programmed cycle stop. The ability to swap tools
quickly and easily had existed on many previous machines. All
turret and capstan lathes, for example, had tools that were
brought in a sequence into the cutting zone. The main operational difference between machining-center tooling and lathe tooling was that lathe tooling was accurately preset on the machine
(generally in combination with turret slide deadstops) whereas
the tooling used on machining centers was preset off the machine (generally in a presetting unit) in advance of its being
needed on the machine. As with the off-line preparation of program tapes, the use of preset tooling was a necessary prereq-
uisite of reducing setup times to a minimum.

A means of reducing setup times further which started to be
used some time after the coming of ATCs was the changeable machine worktable. These enable the next part to be set up on one
worktable while the current part is being machined. At the end
of the machining cycle, the worktables are swapped and changeover times between parts are thus reduced to a few seconds.

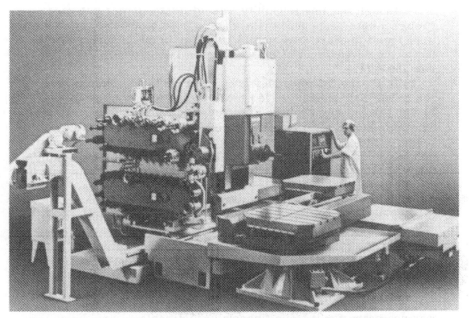

Fig. 2.4 A horizontal machining center with twin exchangeable worktables. (Courtesy Cincinnati Milacron.)

The last productivity-increasing feature of machining centers to be highlighted is the indexing worktable. These enable machining to be carried out on all sides of a part at a single setup and thus reduce the number of setups (and operations) needed on a part, with the benefits already discussed. This is only easily practicable with horizontal spindle machining centers. The productivity benefits of using ATCs, changeable worktables, and indexing worktables combined with numerical control made the horizontal spindle machining center the backbone around which FMS developed. HMCs will be considered again in Chapter 4 with the other machines of FMS. Developments in control systems and computers were, however, equally important.

The 1960s saw NC applied primarily to drilling machines, then to milling and boring machines, and subsequently to machining centers. NC lathes were also available, but as turning is generally a simpler, shorter-cycle-time operation, NC was not seen as offering financial advantages over semiautomatic and automatic lathes. It must be remembered that at this time, NC machines often cost three times as much as conventional machines,

and although machine reliability was adequate, it was not good.
Problems were particularly experienced with tape readers, which
were found to cause more errors than any other feature. This
led to the development of direct numerical control (DNC) in the
late 1960s [12].

In DNC, the NC controller is coupled up to a computer and
the coded instructions are fed block by block from the computer
into the machine controller. The tape reader is thus bypassed.
The approach is sometimes also referred to as "behind the tape
reader" (btr). The DNC computer has storage capacity for many
programs and can be connected to a number of machines, supplying them all with program code, block by block. DNC was not
the first application of computers to controlling machines. In
1966, the first NC machine controlled by a computer was produced, although the computer concerned, like all computers at
the time, was a sizeable cabinet of electronics and very expensive. In the 1960s, the cost of NC controllers often made the
price of an NC machine three times that of a conventional machine, and the use of a dedicated computer to drive a machine
was even more expensive. DNC was a means of spreading the
cost of a computer across a large number of machines, and this
was necessary for it to be financially justifiable.

This brings this brief history of NC up to the date when the
early FMS were being planned. From 1966 onward, how NC developed significantly influenced how FMS developed.

The next milestone in control system development occurred in
1971, when the first machine tool controlled by a minicomputer
came onto the market. The minicomputer was small enough to fit
into the base of existing NC control cabinets and was no more expensive than a "hard-wired" controller. This led to most manufacturers quickly changing over to what became known as computer numerical control (CNC). New hard-wired controllers soon
became a thing of the past because CNC offered so many advantages. In particular, it could store many programs and thus
avoid tape-reader problems, just as DNC had been designed to
do. A tape reader was still needed to read a program the first
time but it was not subsequently needed to reread the tape for
every part produced. CNC also presented the possibility of controllers being linked to and communicating with other computers.
This was an important element in building integrated systems.

CNC temporarily made DNC redundant for new machines but
not for existing older hard-wired controllers. The memories on
CNC controllers were, however, a finite size and could not store

programs for a large number of parts. So it was not too long before CNC controllers were linked to large central computers to take advantage of larger memories for storing programs. The initials DNC continued to be used to describe the arrangement, but they were now to stand for "distributed numerical control." Data no longer needed to be transmitted block by block; rather large sections of programs or complete programs could be transmitted at a time from the DNC computer to a CNC controller. Maintaining the same initials while the older type of DNC remains in existence alongside the new has caused confusion and misunderstanding, but this problem will disappear in time. It should be pointed out that most NC/CNC controllers are not coupled up to central computers but operate as stand-alone controllers. However, almost all controllers in FMS operate under DNC.

2.4.3 Computer Managed Manufacturing

From the 1950s/60s, mainframe computers were applied to the complex logistical problems of planning and controlling batch production. It can be appreciated that the procedures of batch manufacturing should be a ready application for data processing because the control of batch production requires manipulating large amounts of data. Large programs were written and were marketed, mainly by the suppliers of mainframe computers. In many companies, these programs were not successfully applied because the structure of the program did not suit the organization of the company to which it was applied. The company thus had to change its procedures to suit the program, and this was not easy or necessarily beneficial. Some companies found themselves trying to run a manual system and a computer-based system alongside one another during long transition periods, with many resulting difficulties.

One key thing these programs did not do was to integrate the organizational system in the company with the manufacturing system. It was the paperwork systems that were computerized, but only in one direction — from the computer to the shop floor. Little, if any, change took place in the manufacturing system on the shop floor, and no computerization of the information flow from the shop floor back to the computer occurred. Although this cannot be considered a criticism of these computer-based control systems, it does in retrospect help to explain why many of the applications failed to improve the logistics of batch manufacture as much as was hoped.

2.5 REVIEW

This chapter has briefly traced the history of the developments of manufacturing systems up to the 1960s, when the forerunners of today's FMS were under development. The origins of specialized machine functions have been described and shown to be highly beneficial for mass production because with machines like transfer machines, each station of a machine can be designed to carry out a single operation, and only that operation. This gives a very cost-effective design. If only one axis of motion and one spindle speed is required at a particular station, then that is all that is provided and the design can be optimized on the operation to be performed. Furthermore, the linking of a series of single-function, equal-cycle-time stations by some form of work-handling guarantees to minimize part-handling time while maximizing machine and tooling utilization and productive efficiency.

Fixed automation and dedicated machines are of little use in batch manufacture. It has been explained how GT attempts to capture some of the benefits of mass production by creating families of parts with similar manufacturing requirements and cells of machines to machine them. The cell layout minimizes the handling of batches and part similarities minimize machine changeover times in setting up for the next batch of parts. The applications of GT and also of computers were attempts to improve the organizational aspects of batch manufacture.

The advent of numerical control and the development of the machining center provided the first flexible machine around which FMS could develop. "Flexible" is used here in the context given in Chapter 1. Thus it is a machine with stored tools, a stored program, and a loaded worktable which can carry out operations on workpieces in any sequence, having the tools and instructions available to enable a changeover of workpieces to occur without manual intervention. It was this development in the 1960s, together with the growth of NC and subsequently of DNC, together with the concepts of GT and the work-handling approaches of transfer machines, that led to the possibility and realization of FMS. How this happened will be explained in the next chapter.

REFERENCES

1. A. Smith, *An Inquiry into the Nature and Causes of the Wealth of Nations*, Vol. 1, W. Strahan and T. Cadell, London (1776).

2. A. Ure, *The Philosophy of Manufactures*, Charles Knight, London (1835).

3. S. Lilley, *Men, Machines and History*, Lawrence and Wishart, London (1965).

4. J. W. Oliver, *History of American Technology*, Ronald Press, New York (1956).

5. A. F. Burstall, *A History of Mechanical Engineering*, Faber and Faber, London (1963).

6. C. S. George, Jr., *A History of Management Thought*, Prentice-Hall, Englewood Cliffs, NJ (1972).

7. R. Wild, *Mass-Production Management*, Wiley, London (1972).

8. Anon, "Quick-change transfer line machines 13 large components," *Metalworking Production, October*: 120—125 (1973).

9. Anon, "Manufacturing technology — A challenge to improved productivity," *Report to the Congress by the Comptroller General of the United States*, US General Accounting Office, Washington, DC (1976).

10. C. F. Carter, Jr., "Trends in machine tool development and application," *Proceedings of 2nd International Conference on Product Development and Manufacturing, University of Strathclyde*, MacDonald, London, pp. 125-141 (1972).

11. S. P. Mitrofanov, *Scientific Principles of Group Technology* (English translation), National Lending Library for Science and Technology (now The British Library) Boston Spa (1966).

12. J. E. Sandford, "Direct NC speeds programming," Iron Age, 201, 10, March 7, 111—112 (1968).

3

Flexible Manufacturing Systems
—Their Development and Benefits

3.1 PIONEERING INTEGRATED SYSTEMS

The title of this section refers to pioneering integrated systems, but a qualification should be added that only systems for high volume and batch production are included. This qualification is important because transfer lines are integrated systems and they were pioneered 50 years before the systems now to be described.

In the last chapter, the major benefits of using transfer lines and link lines to machine parts were shown to be high utilization for both machines and tooling, low manning levels, and an inherently organized flow of work. The key to these benefits is integration of the work-handling with the machining operations and balancing of the lines by virtue of their design. It was further explained how the use of GT cells was designed to bring the benefits of higher utilizations and integrated work handling (though by manual means) through from mass production to batch production. The development of NC was shown to offer a further means of improving batch production efficiency, while the introduction of DNC started the process of linking machines into larger computer-controlled systems.

The designers of the early integrated systems for batch production saw the benefits of the integrated work handling of transfer and link lines, of the cell approach of GT, and of the emerging
capabilities of NC machines, and they combined the three approaches, seeking their benefits but also looking for other advantages. The advantages and benefits will be summarized and
discussed later in this chapter once examples of pioneering integrated systems and pioneering flexible systems have been discussed and their form and some of their details are appreciated
by the reader.

The systems that will be described have been selected because
together they incorporate almost all the elements of today's FMS
while each illustrates somewhat different aspects of the development of such systems. Where they are lacking in today's terms
is in the scale of integration that is possible with the technology
of modern computers. American, British, and Japanese systems
are described because each contributed to the evolution of ideas
and approaches. Dates are quoted in connection with all the systems described, but these are only approximate. From developing the concepts of a system through to its implementation has
often taken 5 or 6 years and sometimes longer. Once in commission, a system may be further refined. The dates quoted are
the approximate dates of installation, unless this is stated not
to be the case.

3.1.1 The Sundstrand Aviation NC Link Line

One of the first link lines (if not the first link line) that incorporated NC machines and was designed for batch production rather than mass production was the link-line system supplied to
Sundstrand Aviation by the Sundstrand Machine Tool Company
in the mid-1960s. At that time, the Sundstrand Machine Tool
Company and Sundstrand Aviation were all part of the Sundstrand
group. (Since 1977, the Sundstrand Machine Tool Company has
been part of White Consolidated Industries.) The two companies
were (and are) situated about 15 miles apart in Rockford and
Belvidere, Illinois, respectively, and this geographical closeness
was an obvious help in developing an integrated machining system. The system initially installed had eight Omnimill machining
centers and two multispindle drills linked by a powered roller
conveyor system. A plan view of the system layout is shown in
Fig. 3.1.

The workpieces machined by the system were (and are at the
time of writing) aircraft gearbox casings made of cast aluminum
and magnesium alloy. About 70 types were initially produced.

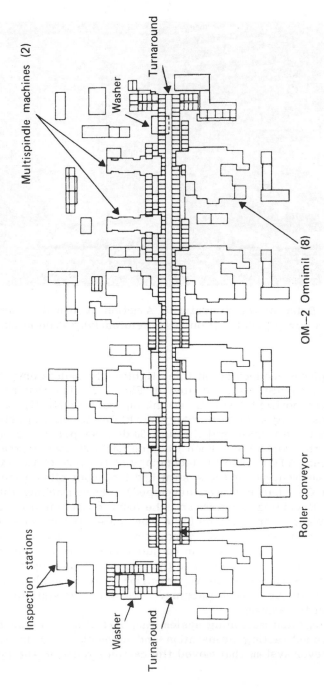

Fig. 3.1 The layout of the Sundstrand Aviation system.

Fig. 3.2 Detail of the Sundstrand Aviation system showing a
loaded and empty pallet. (Courtesy *Metalworking Production*.)

The workpieces are located on thin, lightweight pallets only 40
cm (16 in.) square, as illustrated in Fig. 3.2. The Omnimills
are five-axis horizontal spindle machining centers (HMC), one
of the axes being a vertical rotary worktable which requires
the pallets to be transferred through 90 degrees prior to clamping on to the worktable. Each machine has a 39-tool automatic
tool changer (ATC). The NC technology at the time of the system's development meant the system had tape-controlled machines
(DNC and CNC had not been developed). Each machine thus
needs the machining program and the tools to machine a particular workpiece prior to machining a workpiece. It can thus only
effectively be used as a batch production system. The control
logic of the conveyor system, in conjunction with coded pallets,
ensures that the parts only go to the machines tooled up to receive them. Small buffer conveyors prior to each machine allow
two parts to queue at each machine. Two washing stations are
included in the system.
   The integrated machining system was part of a larger system
which included casting preparation and inspection and an overhead conveyor system that moved the castings between the system

and other processes [1,2]. The machining system with its 10 NC machines replaced 100 conventional machines that used a large variety of specialized jigs and tools. The Sundstrand Aviation system is believed to be the first system that integrated NC machines and work handling in the form of link line. It was a batch manufacturing system, however, and does not have the flexibility to be classed as an FMS. The Sundstrand Aviation system is in operation today in much the same form as it was originally installed. Changes and developments have been incorporated in it and more are planned, including replacing the NC controllers, which have become increasingly out of date and difficult to maintain.

Three or four years later, Sundstrand Machine Tool supplied a system with similar features to Ingersoll Rand at Roanoke [3]. The NC controllers in this case were linked by direct numerical control (DNC). The system had six machining centers, two NC drills, and a roller conveyor system but handled somewhat larger components, made primarily of cast iron. As with the Sundstrand Aviation system, a large number of different parts and restricted tooling capacity at any machine meant that one machine handled one part type at a time in a batch mode of operation.

3.1.2 The Herbert Ingersoll System

The Herbert Ingersoll company at Daventry, England, was set up as a joint venture by two machine tool manufacturers — Alfred Herbert of Britain, at one time the world's largest machine tool manufacturer, and the Ingersoll Milling Machine Co. of the United States. The company was set up in the mid-1960s to produce special machine tools and systems for the world market. The aspect of the Daventry factory of interest is the machining system that was installed in its light machining area in about 1968. The system is illustrated in Fig. 3.3.

The workshop layout is significant because a large number of mainly conventional machine tools were linked together in a systematic and relatively low-cost way [4]. The system consisted of two subsystems, one for machining turned parts and gears (in the lower left of the figure) and one mainly for machining prismatic parts (in the upper right). The prismatic section did have some NC machines, as can be seen. The two "halves" of the system were each served by a control center (as illustrated in Fig. 3.3) and had their machines linked to the control centers by outward and return roller conveyors. Workpieces, tooling, drawings, and instructions were supplied on cafeterialike trays to

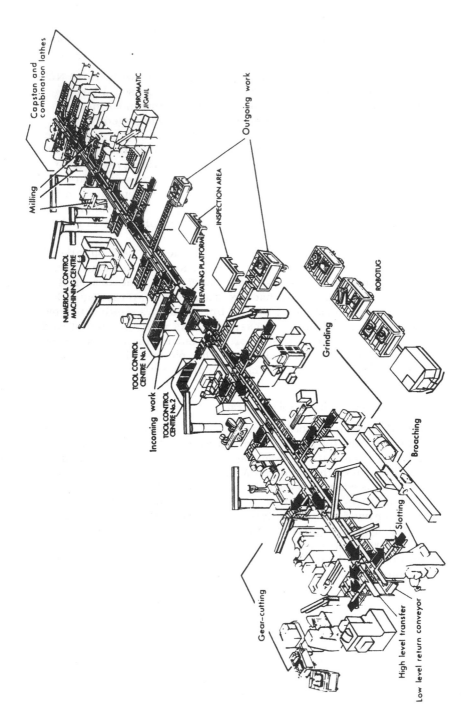

Fig. 3.3 The Herbert Ingersoll system. (Courtesy *Metalworking Production*.)

Capstan and combination lathes

Milling

SPIROMATIC JIGMIL

Outgoing work

NUMERICAL CONTROL MACHINING CENTRE

INSPECTION AREA

TOOL CONTROL CENTRE No.1

Incoming work

TOOL CONTROL CENTRE No.2

ELEVATING PLATFORM

Grinding

ROBOTUG

Broaching

Gear-cutting

Slotting

High level transfer

Low level return conveyor

the operators at each machine via the conveyor — the outward conveyor being mounted above the return conveyor. Figure 3.4 illustrates one of the conveyor spurs to a machine. The system was controlled and scheduled by a foreman in the control center who had to ensure the machine operators were kept supplied with all their needs in advance of a need arising. A telephone link was provided between each operator and the control center to

Fig. 3.4 Detail of the Herbert Ingersoll system. (Courtesy *Metalworking Production.*)

ensure rapid and easy communication. Each half of the system
was similar in concept to a GT cell except that the order and
number of machines visited by any workpiece were controlled
from the control centers.

It is of interest to note that the system was visited by a
"robotug" or wire-guided vehicle (WGV), which towed a series
of carts between 14 work areas within the factory. The robotug
delivered parts to and collected others from the system, with operators manually transferring parts on and off the carts. The
system thus included two forms of handling system for parts.
A similar part-to-machine conveyor system was until recently in
operation in Ingersoll's plant in Rockford, Illinois.

3.2 PIONEERING FLEXIBLE SYSTEMS

The two systems just described were integrated systems; they
cannot quite be described as flexible systems. They were conceived as batch manufacturing systems and had to operate as
such because they were not designed with sufficient flexibility
to machine parts automatically in any sequence rather than in
batches. It is this ability that primarily distinguishes a flexible system from other types of integrated system. Any programmed machine can machine parts automatically, but automatic
machining in any sequence implies the machines can switch part
types automatically. (This does not imply unmanned operation
of the system.) The two systems just described had machine operators who had to be involved in any change of a part.

To be described as flexible requires a system to have the following three features (among others that will be considered later).

1. A means of reprogramming machines automatically to machine
   different workpieces
2. A means of having all the tools necessary to machine the workpieces available at the machines
3. A means of automatically transporting the workpieces between
   machines and automatically loading and unloading the machines

The first feature can be provided either by having NC machines
with a DNC-linked computer or by having CNC controllers with sufficient storage capacity to store the required machining programs.
The CNC controllers need a means of being informed of the next
program to run, so they must either be able to read a pallet code

or be linked to a supervisory computer. The early FMS had DNC-linked NC controllers because CNC had not been developed.

The tooling requirement can be met by machining centers with automatic tool changers, and it was the flexibility provided by the tool changer of machining centers that made them the foundation machines of FMS. Other machine types can also change tools; these will be discussed in Chapter 4. Tool changers enable any available tool to be called up by a machining program and to be changed automatically and quickly. However, a tool changer provides only a limited selection of tools, and the variety of workpieces that can be machined is thus limited by the tools available. It is here that the GT enters the picture. By selecting a GT family of workpieces for an FMS to machine, it is possible to find a range of workpieces that can be accepted by the system.

The first pioneering flexible system to be described, the Molins System 24, achieved the design features specified, but not in the ways described. As far as the DNC link was concerned, the system was under development in the 1960s and it preceded the development of DNC. It adopted a different and ingenious approach. Many of its other design features also have a novelty of approach, making it a step change in system design.

3.2.1 The Molins System 24

Molins are a major manufacturer of cigarette making and packaging machinery. The speed of production of cigarettes is very high — they appear to pour out of the machines. In the early 1960s, Molins wished to expand their output and so they reviewed their manufacturing operation from all sides, including their use of floor space (they had space restrictions) and their use of labor (they had difficulty in getting staff to work shifts). The result of their review was a radically new concept of a manufacturing system that included many other novel ideas as well.

One aspect of their novel approach was that they decided they would make their cigarette machines from light alloy rather than ferrous materials because the machining rate for light alloy is almost unlimited in terms of tool wear. The extra cost in switching from ferrous materials was seen as being more than balanced by the potential savings in manufacturing cost through the use of very high cutting speeds. This change in material led to new concepts for the machine tools to cut it. In reviewing the machine tools on the market, the Molins' engineers could not find an NC machine that would cut aluminum and light alloys at the speeds they wished to use, so they decided they would design their own machines. The machines had many innovative features, but only those relating to the configuration and operation of the manufacturing system will be presented here. Their manufacturing system ideas found final expression as the Molins System 24.

A model of the layout of the proposed system is shown in Fig. 3.5. The system was named the System 24 because it was designed to operate 24 hours a day. This may seem unremarkable

Fig. 3.5 The Molins System 24. (Courtesy *Metalworking Production*.)

and even strange, as most high production systems are planned
to be used 24 hours a day. The reasons for emphasizing this
feature will, however, become apparent. The system was to consist of a line of special NC tape-driven machines. The machines
resembled horizontal spindle machining centers except that there
were single- and twin-spindle machines and the high-speed horizontal spindles were mounted on two-axis slides in front of the
machine column where other horizontal spindle machining centers
have their worktable. The workpieces were mounted on pallets
located on a vertical worktable on the front of the column, where
most horizontal machining centers have their spindle. This configuration facilitated swarf clearance. The pallets were only 32
cm (13 in.) square and were slid into place on rails from one
side of the column. The workpieces were generally machined
from solid slabs of aluminum which were screwed to suitable
threaded holes on the pallets.

The machines did not have automatic tool changers and a
carousel of tools, but rather had a changeable magazine or tooling pallet that carried 14 tools. The magazines were mounted on
the machine column above the workpiece pallet and could be
changed in a similar manner to the workpiece pallet. The magazines were exchanged into a unit that could hold five magazines
in total (an automatic magazine changer). Thus, the machine
could access any of 70 tools when machining a part. On the machine, the tools were mounted horizontally in the pockets of the
magazine with their shanks projecting toward the spindles. Tool
changing was done directly by the spindle swapping tools in and
out of the magazine.

The machine control programs were held on magnetic tapes in
cassettes. The cassettes for each machine for each day's production were held in the central control room in a specially designed
container. The appropriate cassette for a given workpiece could
be accessed under computer control and read in the control room
by an extension of the individual machines' controllers.

Alongside the line of machines was a double-sided computer-
controlled automatic storage and retrieval unit (AS/RS), which
stored the palletized workpieces and the magazines of tools. On
the other side of the AS/RS to the machines was a series of work
benches for fixturing and unfixturing the workpieces from the
pallets and for tooling preparation. The AS/RS consisted of a
large number of storage racks and two traveling conveying units
termed MOLACs (Molins On-line Automatic Conveyor). The
MOLACs were mounted on both sides of the AS/RS and could
access any of the racks. The MOLAC on the machines' side

transferred parts and tools to the machines, the final transfer of pallets from the MOLAC to a machine occurring via short rails, the rails of the tool magazine being above those of the workpiece pallets. The MOLAC on the preparation side of the racks serviced the workers at the preparation benches.

The difficulty of getting workers to work shifts was to be overcome by providing sufficient storage space in the AS/RS to keep the machines supplied with parts and tooling during the second and third shifts, with those involved in workpiece and tooling preparation working only the day shifts. The system was thus to work almost unattended for two shifts. This feature of its almost unmanned operation explains the "24" of the System 24. The average cycle time for the workpieces was 15 minutes, so storage space for 72 pairs of pallets per machine was required for 18 hours of overnight operation.

The design of this system and its machines is of interest for many reasons. The system is fully described by Williamson [5,6]. In terms of the development of manufacturing systems, it has the following features of interest; some of which (though rarely all) may be found in today's flexible systems:

1. The machines were to be supplied automatically with work-
    pieces and tooling.
2. Workpiece loading and tooling preparation were separated from
    the machines.
3. The system scheduling through the MOLACs was under central
    computer control.
4. The system was to be manned for the day shift and incorpor-
    ated enough storage capacity in the AS/RS to operate un-
    manned in the preparation area for the second and third
    shifts.
5. The machines had tool magazine swapping rather than single
    tool changing.
6. The AS/RS conveying unit also carried out workpiece/pallet
    handling for the machines.

It is interesting to note that Molins designed specialized machines for the system. They had two versions of a twin-spindle three-axis machine and a six-axis machine with a single spindle. One version of the three-axis machine was primarily a very high-speed milling machine; the second version was designed for drilling, boring, and hole making. The six-axis machine was designed to cater for parts with complex geometry. This approach with machine specialization was taken to ensure maximum metal-cutting rates and productive output from the machines. The requirement to handle the parts more frequently and the restricted machine flexibility were accepted.

Just as the system incorporated many novel features that are now fairly common, so did the machines. For example, the machines had spindle torque monitoring as a means of in-cycle broken-tool detection. They also incorporated on-machine tool setting, and this was allied to a further means of broken-tool detection. If the tool could not be set close to the programmed setting position, it was assumed to be broken.

Molins realized that the machines they had developed had commercial application for other companies using light alloy components. They actively marketed them and sold a number to customers who were mostly in or associated with the aircraft industry. Molins themselves never fully implemented the System 24, for reasons discussed by Foster [7]. The philosophy behind the system was, however, published, and its system concepts and ideas were seen both as challenging the then-current thinking and as pointing a way forward.

3.2.2 The Allis Chalmers System

Allis Chalmers manufactures tractors (among other products), and in the late 1960s, the firm realized they had to update their tractors to newer, larger designs if they were to stay competitive in the tractor business. A new product range was designed, and this had to be manufactured, but because it was new, it was realized it was likely to undergo certain design changes during its first two or three years. It was also accepted that the life of the product might be only five years and a substantial redesign might then be necessary to react to changing market requirements.

A manufacturing system was required to machine the major gearbox and transmission castings and their options. Allis Chalmers had used transfer lines to produce such components in the past, but the possibility of engineering changes or a complete redesign made this seem a highly risky approach to take. Allis Chalmers was looking for the flexibility possible from NC machines but the manning levels achievable with transfer lines. A production requirement in terms of parts and tooling was given to a number of potential suppliers. Approaches using transfer lines, stand-alone NC machines, and a flexible integrated system were all considered. A collaborative contract for a flexible system was eventually awarded to Kearney and Trecker (K&T) because (1) they seemed really interested, (2) it appeared to be the lower

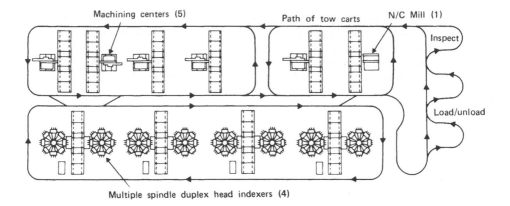

Fig. 3.6 The layout of the Allis Chalmers FMS.

total cost approach, and (3) they were located less than 5 miles
away and could easily use the Allis Chalmers plant as a development laboratory. Contract costs were shared. The total machining requirements could also have been met by 13 stand-alone
NC machines but at a greater capital and labor cost than the system chosen.

The system installed in 1972/73 consisted of six K&T moduline
machining centers, one of which was to be a rough milling machine, and four duplex head indexers [8]. The system layout is
shown in Fig. 3.6. The lines surrounding the machines indicate
the paths of the automated tow cart work-handling system. Part
of a smaller tow cart system is illustrated in Fig. 3.7. The workpieces can be seen mounted in fixtures, which in turn are mounted on pallets that are carried on the carts. The carts are simple
devices having four castored trailing wheels and a towpin at the
front, which engages through a slot in the floor with continuously driven chains. The carts are controlled from the floor by
having the pin engaged or disengaged. Disengagement is accomplished by having a "shark's fin" cam raised from the floor.
If the cart is stopped at a machine shuttle loader, it is positively
pushed against a stop to locate it accurately for pallet transfer
to a machine shuttle loader. Tow carts and various other types
of work-handling device are described and compared in Chapter 5.

Fig. 3.7 Loaded tow carts. (Courtesy Kearney and Trecker Corporation.)

The Allis-Chalmers machines are served by similar five-posi-
tion linear shuttle loaders. The central position of the shuttle is
the machine worktable position, and there are four waiting positions, two on each side of the worktable. The shuttles pull the
pallets off the carts at one end of the loader and push them onto
an empty cart at the other. The carts are thus completely passive transporters. The provision of four waiting positions, two
on each side of the machines' worktable position, enables the machines to operate independently of the work-handling system. It
additionally provides some system storage.
    The machine combination of machining centers and head index-
ers is interesting. Although machining centers can be considered
completely flexible machines, the unit heads of a head indexer are

a form of both fixed automation and flexible automation. Each
unit head is fully tooled up to produce a given pattern of holes
of predetermined diameters or given bore sizes with fixed relationships. The ability to index the head gives flexibility, but the degree of flexibility is limited by the number of heads on the indexer. The machines of FMS are considered further in Chapter 4.

By 1970, the DNC approach had been developed and the Allis Chalmers system was installed with DNC control. The machining centers could thus machine different workpieces in any
sequence under command from the system controller (as long as
the machines had the necessary tools, and this was checked by
the controller). The combination of features that gives the system the ability to switch workpieces under computer control without a delay for a changed setup makes the Allis Chalmers system
a truly flexible system, and because the Molins system was not
fully implemented, the Allis Chalmers system can be considered
the first FMS.

The system was originally designed to machine a small family
of transmission and clutch housing of sizes up to 875 mm × 500
mm × 500 mm (35" × 20" × 20") weighing up to 275 kg (600 lb).
In the years of operation, the parts that have been produced
have changed and the part mix has changed. For some parts,
new indexer heads have been required; for others, more tooling for the machining centers. The volume requirement for one
part increased sufficiently that it was taken off the system and
put onto a transfer line. During its life, volumes of parts produced per year by the system have ranged from 13,000 to 28,000;
the higher volumes involving continuous operation on a three-
shift basis. In the early 1970s there was a large demand for
tractors for the U.S. market because President Nixon had laid
down the policy that the United States should be the breadbasket
of the world. During its early life, the system really proved its
worth as new parts and products were introduced.

3.2.3 The Fujitsu Fanuc Hino Plant System

The integrated and flexible systems described so far were all designed to machine prismatic (i.e., nonturned) parts, and all used
various forms of horizontal spindle machining center either exclu-
sively or in combination with other machines. In Chapter 2, it was
pointed out that NC was developed and applied to the machining
of prismatic parts well before it was considered economically advantageous to use NC for turning. It was not that the NC turning machines were not on the market; they were. It was the fact

that there were other, much cheaper automatic and semiautomatic lathes available to produce turned parts. It is not surprising, therefore, to find that integrated flexible systems for
turned parts did not appear until slightly after those for prismatic parts.

The first such system known to the authors was the one built
by Fujitsu Fanuc and installed in their Hino plant [9]. The system is notable not only because it incorporated NC lathes, but
also because the work handling was done by a gantry-mounted
industrial robot. The system is shown diagrammatically in plan
view in Fig. 3.8 It consisted of eight NC lathes under DNC control and an industrial robot mounted upside down above the machines on a gantry on rails, very much like a workshop crane.
Beside each lathe was a carousel feeder carrying workpieces.
The robot's function was to transfer workpieces between the carousel feeders and the lathes and to turn the workpieces around
between first and second operations. The eight lathes were at-
tended by one man who kept the carousel feeders replenished.

Implementation of this system required the development of a
special robot gripper to accept workpieces of a variety of shapes
together with interlocks and interfaces between the robot, the
carousel feeders, and the lathe chuck and lathe doors. A modern development with similar carousel feeders but with individual
pick-and-place robots mounted on the side of the lathes is shown
in Fig. 3.9.

It is debatable whether the Fanuc system should be included
here as a flexible system or in Section 3.1 as an integrated system. The system was not designed for flexible operation and had
a restricted range of tooling. However, because of its DNC system and other mechanical and control features, it is believed flexible operation was perfectly possible and so it is included as a
flexible system. Many DNC systems were installed in Japan in
the early 1970s, and this soon led to the development of forms of
FMS.

3.2.4 Cincinnati Milling Machine Company's VMMS

During the 1960s, the Cincinnati Milling Machine Company (now
Cincinnati Milacron) devoted considerable time to developing
concepts for advanced batch manufacturing systems [10]. As
part of this program, they designed conveyor systems to integrate with their machining centers using loop layouts so that
parts could take variable routes to visit any machine within a
system. An example of an early system and the concept is shown

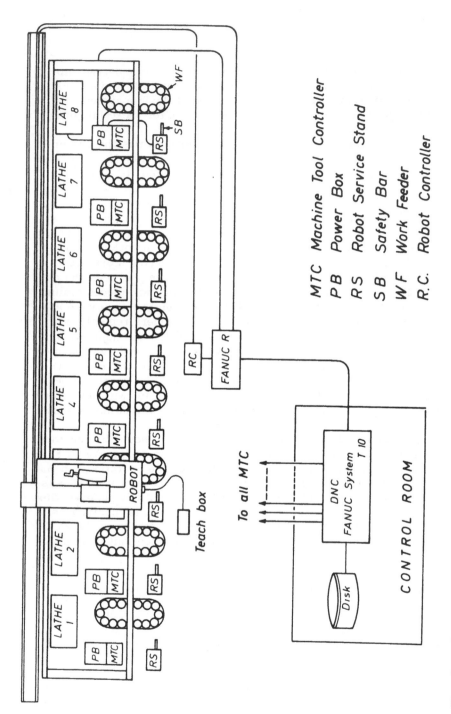

Fig. 3.8 The layout of the Fujitsu Fanuc system.

MTC Machine Tool Controller
PB Power Box
RS Robot Service Stand
SB Safety Bar
WF Work Feeder
R.C. Robot Controller

Fig. 3.9 A lathe with carousel feeder unit and automated loading. (Courtesy Yamazaki Machinery UK Ltd.)

in Fig. 3.10. This system was designed for Borg Warner for their Letchworth, England, factory for machining housings for automatic transmissions. The system was installed in 1968 and at that time had only unpowered conveyors and NC-controlled machines [11].

In the early 1970s, Cincinnati developed a prototype FMS that they termed a "variable mission manufacturing system" (VMMS). The prototype system was built and run in their Cincinnati plant and consisted of three horizontal spindle NC machines and a conveyor system for work handling. The machines were a powerful rough milling machine, a head changer that was primarily used for drilling patterns of holes, and a drilling/boring machine. The system layout is shown diagrammatically in Fig. 3.11. The remarkable feature of the system was that pallet-mounted workpieces were both machined and transported to the machines upside down, the pallets being mounted underneath the conveyor. The workpieces

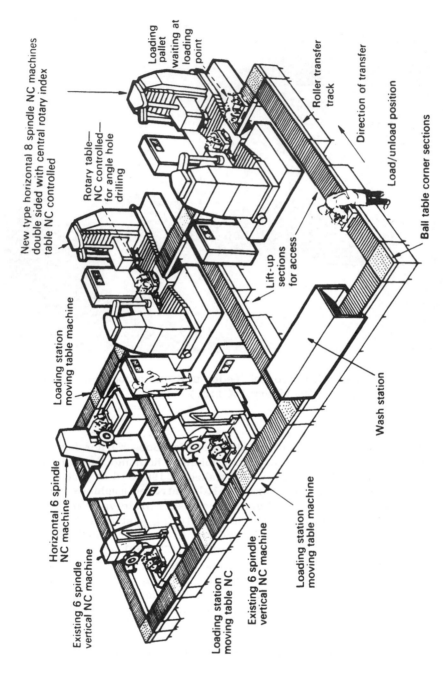

New type horizontal 8 spindle NC machines double sided with central rotary index table NC controlled

Loading pallet waiting at loading point

Roller transfer track

Direction of transfer

Load/unload position

Ball table corner sections

Rotary table—NC controlled for angle hole drilling

Lift-up sections for access

Wash station

Loading station moving table machine

Horizontal 6 spindle NC machine

Existing 6 spindle vertical NC machine

Loading station moving table NC

Existing 6 spindle vertical NC machine

Loading station moving table machine

Fig. 3.10 The Borg Warner system. (Courtesy *Metalworking Production*.)

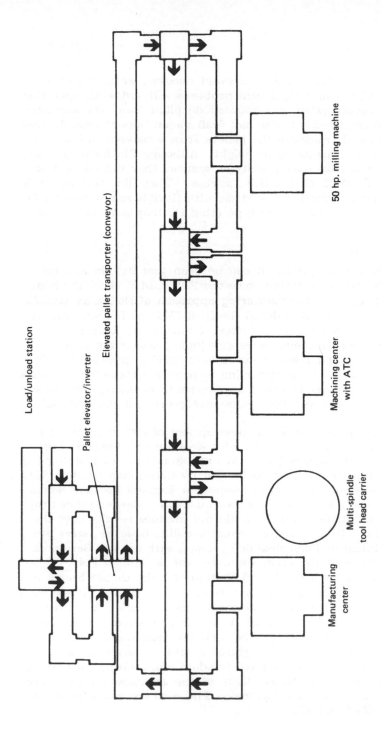

Load/unload station

Pallet elevator/inverter

Elevated pallet transporter (conveyor)

50 h.p. milling machine

Machining center
with ATC

Multi-spindle
tool head carrier

Manufacturing
center

Fig. 3.11 The layout of the Cincinnati Variable Mission Manufacturing System.

were so orientated to facilitate swarf removal, which it was anticipated might cause significant problems with automatic operation.
The conveyor system was mounted on pillars and was substantially proportioned. It also had "rail wagon tipper"-type devices
to invert and transpose the pallets from a conventional load/un-
load height to the conveyor height. Although Cincinnati undoubt-
edly learned from their prototype system, they did not sell any
"inverted" VMMS on the U.S. market. They did, however, sell
integrated manufacturing systems with flexible manufacturing facilities of the fixed-mission type (that is, they were flow-lines).

3.2.5 Review

From these descriptions, it can be seen that Britain and the
United States both started down the FMS road at about the same
time. In Britain, the pioneering approach of Molins System 24
could probably be considered the first FMS had it been fully implemented. In the United States, Sundstrand, Kearney and
Trecker, and Cincinnati were developing similar approaches,
with Kearney and Trecker and Allis Chalmers achieving the first
flexible system. It is interesting to note that these four systems
were all effectively in-house developments involving very close
collaboration between the system designers and the eventual
users.

In the United States, the development of FMS continued on a
slow scale through the 1970s, with Kearney and Trecker supplying systems to Rockwell and Avco and Sundstrand supplying the
first of a number of systems bought by Caterpillar Tractor [12].
The Molins work was not developed in England because Molins
was not really in the machine tool business. Certain other projects were initiated, but these did not produce results until the
1980s [13]. Other European countries did, however, start producing integrated and flexible systems, with the two Germanys
leading the way with other countries not far behind [14].

The Japanese had followed Britain and the United States in
developing integrated systems, concentrating on DNC integration.
The 1970s saw a range of developments. Okuma, for example,
developed their Parts Centre 1, which processed turned parts
from bar stock through a lathe, grinder, gear cutter, and induc-
tion hardener within an integrated system. Many other Japanese
companies increasingly started to adopt and implement a systems
approach to manufacture. Many of the Japanese systems were
small cells of very few machines. These do not really qualify to
be described as FMS.

British and American interest in FMS "took off" in the late
1970s and early 1980s for basically the same reason — the threat
from the Japanese. Industrialists on both sides of the Atlantic
realized that Japanese imports and their penetration of other
world markets could only be contained by producing products
at better quality and lower cost, and this generally meant investing in new technology. The sums committed by some companies to industrial modernization have been substantial and,
where appropriate, these have included FMS. The number of
companies able to supply FMS also increased as machine tool companies realized that their ability to sell machines would depend
increasingly on their ability to sell systems of machines.

3.3 FLEXIBILITY EXAMINED

The systems described have incorporated a large variety of machine tools and work-handling hardware and have illustrated the
diversity that is possible. Later chapters will consider the hardware and software of flexible systems in more detail and will illustrate that further combinations of machines and handling equipment are possible.

To reinforce what distinguishes a flexible system from what
has been termed an integrated system, one further system will
now be discussed. This is not a pioneering system but one that
was implemented in 1984/85. The layout of the system is shown
in Fig. 3.12; it is radically different from the form of those previously discussed (which is one of the reasons for considering
it). The system is to machine gears for printing machinery. The
machines for producing the gears are linked together into a system by the use of two industrial robots and a transfer unit. The
machines involved are two CNC chucking lathes for turning the
gear blanks, typically through a first and second operation; a
gear hobber for cutting the gear teeth, fed automatically by a
carousel loader mounted around its tailstock; a gear deburring
machine; an CNC drill for producing any holes required for
mounting the gears; a washing and a gauging machine to check
the gears. The range and functions of the robots and transfer
arm should be clear from the figure. The gear blanks and cut
gears are handled individually by these devices. The operation
of the system, the industrial robots, the transfer unit, and the
sequencing of parts are under central supervisory computer control, and the system can operate automatically and unattended.
Is the system a flexible system, however?

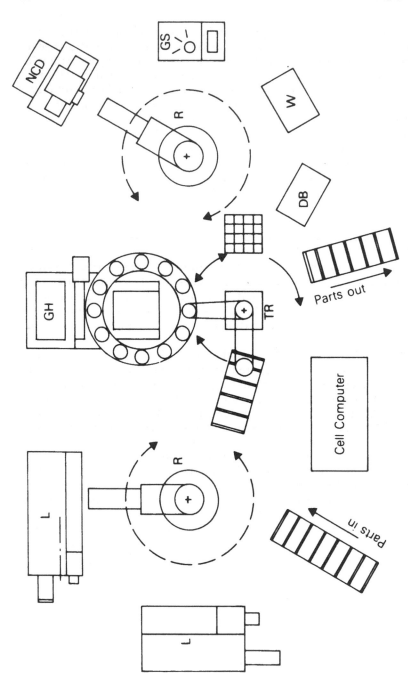

Fig. 3.12 The layout of a gear-cutting cell. L, NC lathes; GH, gear hobber with carousel loader; NCD, NC drill; GS, gauging station; W, wash; DB, deburr; R, robot; TR, transfer arm.

From the information given, neither the hobber nor the deburring machine has been stated as having a CNC controller and so they, and thus the system, can produce only one part type at a time. Because the system cannot switch components automatically, it is not a flexible system.

It might be supposed that the system could be made more flexible if the hobber were a CNC hobber. Such machines are available and they can be programmed to adjust themselves for machining different gear diameters and different numbers of gear teeth. This introduces some flexibility, but not much. For the hobber to be able to machine gears of different modules, a hob-changing hobber would be required together with hob diameter offsets in the controller. Even this only gives slightly more flexibility. The other constraint on the flexibility of the hobber is the hobber fixture, which has to locate the gear blanks very accurately during hobbing. With ingenious design, it would be possible to design a fixture that would accept gear blanks with different bore sizes. Alternatively, mounting flexibility might be achieved by the use of some type of gear bore adaptor that would permit every gear to fit on a standard fixture. Such an adaptor would have to be fitted between turning and hobbing and would need to be a transition fit or better if adequate gear accuracy were to be achieved. This would present considerable technical difficulties as would the alternative of designing a universal fixture. This discussion and example have been used to illustrate that it takes more than a computer, DNC-linked CNC machines, and a work-handling system to give a flexible system. An FMS can only be built around flexible machine tools.

Definitions of a flexible manufacturing system can now be reviewed in the context of the systems described. An FMS might be simply defined as follows: "A manufacturing system is flexible if it is capable of processing a number of different workpieces automatically and in any sequence." The word "automatically" is included in this definition to indicate that the system operations can occur and be changed without human intervention. "Processing" is included for two reasons. First, processing ensures that systems for forging, forming, and other noncutting processes are included within the definition. Second, it suggests more than a single machining operation but less than a complete raw-material-to-finished-part sequence of operations. The definition is not a completely adequate definition, however, because loading/unloading pallets is a manual activity and this is ignored by use of the word "automatic." The definition only starts to be adequate if you already know what an FMS is. A definition that relates more to the system hardware might be phrased as follows:

A number of workstations, comprising computer-controlled machine tools and allied machines, which are capable of automatically carrying out required manufacturing and processing operations on a number of different workpieces, with the workstations being linked by a work-handling system under the
control of a computer that schedules the production and the
movement of parts both between the workstations and between
the workstations and system load/unload stations.

This sentence of 67 words is an adequate definition of the
"what?" of FMS but it totally omits the "why?" To expand the
definition to include the why would make for confusion rather
than clarity. This short diversion into semantics should have
reinforced the point made in Chapter 1 that it is difficult to produce a short and informative definition of FMS.

Some writers on FMS (see, for example, Ref. 15) have felt it
necessary to consider the various types of flexibility available
from a system. These have included the fundamental flexibility
of having a part mix, the ability to change the part mix, the volume mix and the design of parts, and the ability to schedule parts
around various routes within the FMS. Because all FMS offer all
these flexibilities as part of their basic design, there seems little
point in suggesting there are different types of flexibility. However, there are degress of flexibility, and the nonquantitative
scale can be said to range from zero, with a transfer line, to
high, with an FMS of a number of identical machining centers.

The term FMS is now widely known and is often applied to
systems whether they are truly flexible or not. One of the reasons for this is that most company managers, vice-presidents,
and presidents now have some understanding of what FMS are
and what they offer. The concept does not need to be explained.
It is thus far easier for engineers submitting proposals to senior
management for computer-controlled systems to call them flexible
systems rather than to draw the distinction between flexible and
nonflexible. The same is true outside the company. A number
of governments have supported the implementation of FMS through
cash grants or accelerated depreciation allowances. This has resulted in the term FMS being stretched to ensure that any system
resembling an FMS qualifies for a grant.

The gear system just described cannot be called an FMS but
it could be referred to as a computerized manufacturing system
(CMS) or as a computer-integrated manufacturing system (CIMS).
Both these expressions are more general than FMS and hence can

embrace a wider range of system types, as long as a computer is involved somewhere. CIMS emphasizes the computer integrating the system and is thus applicable to systems that may not have integrated work-handling linking machines. It is unfortunate that new and developing technologies suffer from vocabularies and jargon words which also develop in their meaning. One acronym that has established itself with a reasonably specific meaning is CIM — computer integrated manufacturing. This is used to refer to the integration of systems rather than the integration that occurs within systems. Thus an FMS may be part of a larger CIM system. This is considered further in Chapter 6.

Now that the basic elements and some configurations of flexible systems have been introduced, the "why?" of FMS, that is, the reasons for investing in FMS and the benefits to be gained, will be presented.

3.4 THE ADVANTAGES OF FMS

The combination of technologies and approaches to manufacture brought together in FMS offers many advantages. The main occurrence that generally led the companies reviewed to invest in FMS was the introduction of a new product or range of products that required new manufacturing facilities. Companies that are just implementing FMS are also generally using the opportunity presented by the introduction of a new product to invest in new technology and to upgrade their manufacturing facilities. The American companies, such as Allis Chalmers, Rockwell, and Caterpillar Tractor, that pioneered the application of FMS were high-volume manufacturers who selected FMS in preference to transfer lines to ensure both that they could produce their new products quickly and that their production facilities would be able to meet a changed market in the future.

The benefits and advantages of FMS will be quantified where possible in relation both to the systems already mentioned and to systems installed in the early 1980s. The more recent systems include:

The $16-million system supplied to General Electric for its Erie, Pennsylvania, plant by Giddings and Lewis for machining some large motor frames (average size 760 × 760 × 1140 mm and average weight 1133 kg) and some smaller gearboxes.

The system has nine heavy-duty machine tools of four different types (see Chapter 4 and Fig. 4.13).

The system supplied to Hughes Aircraft for its El Segundo, Cali-
fornia, plant by Kearney and Trecker for machining light alloy die castings for aircraft range finders. The system has
nine machining centers and a coordinate measuring machine
linked by tow carts (see Chapter 5 and Fig. 5.6).

The two systems that Mazak built for its own use at the Florence,
Kentucky, machine tool plant. The larger system here is used
to machine beds, bases, and columns, and the smaller system
machines castings for gearboxes, headstocks, turrets, tail-
stocks, and small workpieces.

3.4.1 A Flexible Manufacturing Capability

The inherent flexibility of the machines originally built into an
FMS ensures their ability to accept new workpieces readily. With
new workpieces, investment in some new tooling and fixturing
may be required together with the appropriate machining programs, but little else should be necessary. The Allis Chalmers
FMS has produced a range of different parts during its (continuing) life, although the range of parts produced at any one time
has been small. In contrast to the early systems installed by
high-volume manufacturers, some recently installed systems have
been planned from the outset to machine over 400 different parts.
The concept of flexibility can be considered to have been extended
recently when two existing systems were sold by their first owners, installed in different factories by their new owners, and used
to produce entirely different parts.

3.4.2 High Equipment Utilization

The system computer ensures that a part is only scheduled to a
machine that has the tools to machine the part and will shortly be
available to machine it. The part arrives at a machine already
loaded in its fixture and in advance of the machine being ready
to accept it. The time between parts is only that required to
transfer the pallets and download a new program to the machine
controller. The time between parts is thus minimized and high
equipment utilization can be achieved. As with transfer machines,
a system's design and operation avoids all the typical logistical
problems of batch production which reduce machine utilization.

3.4.3 Reduced Equipment Costs

The high machine utilization possible with FMS (typically between 75% and 90%) gives a greater productive output per machine and hence reduces the number of machines required for a given output compared with stand-alone machines. The reduced cost of the machines required is usually more than enough to cover the extra costs incurred for the work-handling equipment and computers. For example, General Electric reports that its system's nine machines give 38% more machining capacity than the 29 machines they replaced. The K&T Avco system of 11 machines together with two stand-alone machines will reportedly do the work of 67 conventional machines. Hughes ordered its 10 machine FMS in preference to 25 stand-alone CNC machines at 70% the cost of the stand-alone machines.

3.4.4 Reduced Floor Space

Although some FMS have been liberally endowed with floor space, FMS can be very efficient in their use of floor space, particularly if they have conveyor or tow cart work-handling systems. The figures just quoted for the reduced number of machines required to give a certain productive capacity show the possible scale of the reduction in floor space requirements. FMS will always occupy far less than the equivalent space of the stand-alone machines required to give the equivalent productive output. GE quotes a floor space saving of 25% for 38% increased capacity.

3.4.5 Reduced Direct Labor Costs

The level of manning required for an FMS depends on the number of machines in the FMS and the workpiece cycle times. Because part loading and unloading is centralized and one "operator" can look after six or seven machines and the total number of machines required by an FMS is low for a given output, the number of people required to man an FMS is small in comparison with the number required to achieve the same output using stand-alone machines. The savings depend on the company situation and existing manning levels. FMS need full-time maintenance support, however, and may also need a full-time tooling operative. (The manning of systems is dealt with more fully in Chapter 6.) Companies do not usually report this saving as a labor reduction but

rather as an increase in productivity per employee. GE quotes
a 24% improvement here, although it is believed that most systems achieve more than this.

3.4.6 Shortened Lead Times, Reduced WIP, and Improved Market Response

Early FMS users were primarily looking for flexibility. Two linked
benefits they obtained by virtue of the inherent design and operation of FMS were shortened manufacturing lead times and reduced work-in-progress (WIP). These together enable them to
respond to changes in the market quickly without having to
hold large stocks.

Once a part has entered an FMS, it will naturally be processed
through its operations because only very limited-length queues
and backlogs are allowed to build up in a system. A number of
features of the design and control of a system virtually guarantee this. These are: the computer system scheduler; the monitoring of the progress of pallets while they are in a system; the
limited in-process storage; and the fact that all the parts not in
a system are in the vicinity of the loading stations and under the
supervision of the load/unloader.

Large benefits are available here because the one-week-per
operation approach typical of much batch production is eliminated
completely. Even GE, which used to process parts in 16 days
(already a short time period), reduced its throughput time to 16
hours. Mazak reports a reduction of 75% in its in-process inven-
tories and throughput cycle times reduced by over 90%.

3.4.7 The Simplification of Manufacturing and Improved Management Control

The organizational problems of managing batch manufacturing were
discussed in Chapter 2. An FMS has similarities to a GT cell in
that the approach of scheduling a batch to the system or cell
means that it is processed by the system or cell without further
scheduling being required outside the system or cell. In addition, the internal scheduling of an FMS is under computer control. Parts are scheduled individually and processed as parts.
This level of control, together with the availability of computer
aids to scheduling the system, makes the system performance pre-
dictable. Expediters are unnecessary, and the system manager
can interrogate the system computer to find out where any part

is at any time. The reduced number of machines of an FMS also eases the organizational task. FMS thus allow planning certainty, simplified planning, and considerable management control.

3.4.8 Gradual System Degradation

A basic design principle of FMS is to have as many identical machines as possible. All parts will then have alternative routes to follow, and this enhances flexibility. If a machine is temporarily out of commission, there will always be other similar machines or alternative ways to carry out the necessary operations. This is not only flexibility but also a form of system redundancy. Other forms of redundancy will also be included in a system so that a system progressively degrades when a fault occurs, rather than stopping completely. Redundancy is not only catered for by having duplicates of machines, it is also built into the work-handling system and the computer controllers in ways that will be explained. This is a significant advantage over transfer machines or any form of fixed flow-line in which parts can only progress if every part of the system is operational.

3.4.9 High Product Quality

Producing parts on an FMS does not guarantee higher product quality but it provides an environment in which it is possible. This results from the automation features built into systems and from the systems being designed as a whole. The fact that machines are kept working and are thus not passing through large thermal cycles also helps to maintain consistency from part to part. FMS do not necessarily give high product accuracy. This topic will be considered further in Chapter 4 when the machines of FMS are considered, and again in Chapter 6.

3.4.10 Phased Introduction

Most of the benefits considered so far have been benefits of FMS compared with more conventional means of batch manufacture including stand-alone CNC machines. In contrast, the benefit of a phased implementation relates more to comparison with manufacture by transfer line. A transfer line has to be fully commissioned for it to perform at anything like its operating potential. Thus, the complete investment must be made all at once

before any parts are produced and the system starts generating
any earnings.

    FMS can be implemented in phases very successfully as long
as plans have been laid from the beginning for the phased investment. This results directly from the flexibility of machining
centers. The third system supplied by Kearney and Trecker is
an excellent example of a system that was installed in phases.
This was the system installed at Avco Lycoming for machining
engine crankcases. The first phase consisted of 5 stand-alone,
tape-driven, NC machining centers linked by a tow cart work-
handling system. The handling system was operator controlled.
This phase was installed by 1976. The second phase, which was
implemented 2 years later, increased the number of machining
centers by four and also added two simplex and one duplex head
indexers. The tow cart circuit was extended to include the new
machines, and a fixture/pallet load/unload area was added. The
machines were all brought under DNC control and the tow cart
system under complete computer control. These additions made
it a fully flexible system. Plans for a third-phase extension
have been drawn up, though not yet implemented.

3.4.11 Just-in-Time Manufacture

This has been included as the last advantage because it is the
most modern one. However, in many respects it is just a re-
statement of the shortened lead times and the reduced work-in-
progress benefit. Just-in-time (JIT) manufacture is the term
given to the "Kanban" approach originally developed and used
by the Toyota Motor Company [16]. Although JIT is a system
of producing parts and complete products just as they are needed, it is more accurately described as a combination of a scheduling system and a means of minimizing inventory. The minimum
inventory is not held in stores but at the output of every operation. The production of more parts at a particular operation is
controlled by the ordering of parts (or a batch) by the subsequent operation. The Kanban is the reordering document or
card that passes between operations to trigger the supply of
the appropriate number of parts from the previous operation.
Because these parts will be supplied immediately, depleting the
finished inventory held, the Kanban also triggers the production of the same number of parts to make up for the inventory
supplied. To produce these parts, a Kanban ordering the necessary raw material will need to be sent to the appropriate

supplying operation. Thus, a succession of Kanbans moves backward through the operation sequence whenever a finished product
is demanded from assembly. Production is thus controlled by demand or "pull." This is in complete contrast to the push approach
used by conventional scheduling systems. All the paperwork that
accompanies conventional scheduling is replaced by Kanban cards.
The initial establishment of inventory levels at the various operation stages needs careful planning, which can only be done successfully if reasonably accurate forecasts of demand can be made.
This is more practicable with high production volumes than with
smaller, more varied demand, as occurs in batch manufacturing.
Hence the development of JIT by Toyota and its adoption by other
high-volume manufacturers. To be completely effective, the JIT
philosophy should extend through from a final product manufacturer back to suppliers and subcontractors so that inventory is
minimized at all stages.

The philosophy of JIT of minimizing inventory levels and producing parts quickly as and when they are needed is very attractive and relevant to all types of manufacturing, particularly batch
manufacturing. FMS have the ability to produce a wide range of
different parts quickly and they thus fit well into the JIT approach. In addition, it can be stated that without some form of
FMS or flexible manufacturing facility, it is difficult to implement
JIT effectively within a batch manufacturing environment. The
JIT approach has been adopted or is being investigated by an increasing number of companies, and this in turn is highlighting
the benefits of FMS.

3.4.12 Financial Benefit

Although none of the advantages of FMS has been expressed in
financial terms, it should be apparent that many, though not all,
of them could readily be quantified as a financial benefit. Companies are not generally prepared to disclose details of the benefits of their system(s) in financial terms, for understandable reasons. However, some companies have been prepared to disclose
the numerical and percentage improvements in performance they
have gained, and it is these that have been quoted.

The actual dollar value of savings depends on the local conditions of the company, on its state and federal taxation regime
in terms of investment grants and allowances, and on its own particular costing system. These factors vary from company to company. Thus, although dollar values of the savings achieved would

be of interest, the actual dollar value does not have as much meaning as the types of figures that have been quoted.

In gathering material and doing research for this book, many U.S. and some European companies were visited and their engineers interviewed. All expressed satisfaction with their systems. It is also significant that certain companies have invested in more than one FMS. Caterpillar Tractor, for example, has seven systems of various types in the United States and four in Europe, and John Deere has three flexible systems operating. This illustrates their confidence in FMS. Some engineers felt it was unhelpful to quantify the benefits of FMS because only some of the benefits could be quantified. For example, how is the benefit of flexibility quantified, or the improved management control? For some, the flexibility benefit was considered so fundamental to the decision to invest in FMS that to quote other benefits in numerical terms and not the main benefit appeared inconsistent. Having said that, companies have invariably quantified the benefits they seek as best they can and have used these figures in an economic justification. Approaches to the economic justification of systems are covered in Chapter 13.

3.5 DIFFICULTIES WITH FMS

Although there are many benefits to using FMS, certain difficulties are associated with them that should be recognized; these are now reviewed.

FMS are expensive, generally requiring an investment of millions of dollars. Substantial capital resources are required for a company to consider installing an FMS. For smaller companies with restricted capital resources, there are two immediate ways around this difficulty. The first of these is to carry out a phased implementation of a system. This reduces the initial capital outlay and enables the earnings generated from the initial phases to help fund the later ones. A small initial phase also gives a shorter commissioning time and shorter learning curves for company personnel. A positive cash flow is generated more quickly.

The second approach can be termed "gradual integration" — that is, to invest in small flexible manufacturing cells (FMC) or even single-machine modules (FMM) which may subsequently be linked together with a work-handling system. An FMM is a single machining center with a carousel pallet storage system. (Features of FMM will be discussed in Chapter 5.) An FMC involves a minimum of two machines with integrated work handling. The essential elements of a flexible approach can be built with both these configurations, and this facilitates subsequent integration. Integrating machines and work-handling systems is easier than integrating their controllers, however, and controller compatibility should thus be considered carefully if this approach is taken.

FMS are more complex than multistation transfer lines. Commissioning and developing a system takes time, and as every FMS is different and tailored to the requirements of a particular customer, every system poses its own challenges. Not all existing purchasers have approached this realistically. Commissioning a large system can take up to 18 months, and running the system up to a high performance level can take another 18 months.

Implementing a first FMS is a step change for a company at all levels. It requires a highly supportive and knowledgeable management and an adaptable workforce that has been involved in developing the system requirements. It requires maintenance personnel with varied skills. Its effect may be widespread. Careful planning and communication are as necessary on the human side of the company as on the technical details of the system.

Having reviewed features of a number of systems and considered the benefits and difficulties of FMS, attention will now be turned to the detailed system elements that have to be chosen as part of the design process.

REFERENCES

1. B. C. Broscheer, "The NC plant goes to work," *American Machinist*, Oct. 23, 138--149 (1967).

2. B. C. Broscheer, "N.C. linked-line goes to work," *Metalworking Production*, July 24, 41--47 (1968).

3. J. E. Standford, "DNC lines link cutting to a new future," *Iron Age*, Oct. 26, 46--49 (1972).

4. W. A. Hawkins, "High productivity machine-tool factory," *Metalworking Production*, April 10, 35--39 (1968).

5. D. T. N. Williamson, "Molins System 24 — A new concept of manufacture," *Machinery and Production Engineering*, Sept. 13, 544--555 and Oct. 25, 852--863 (1967).

6.   D. T. N. Williamson, "System 24 — A new concept of manufacture," in *Proceedings of 8th Int. M.T.D.R. Conference, UMIST*, Manchester, Pergamon Press, Oxford, London,
     pp. 327-376 (1967).

7.   G. Foster, "The making of Molins," *Management Today,
     March*, 66-69, 110, 112 (1976).

8.   Anon., "Flexible manufacturing system fulfills tractor maker's
     needs," *American Machinist, 119*:45-64, June 15 (1975).

9.   A. W. Astrop, "DNC used to manufacture NC equipment,"
     Machinery and Production Engineering, June 8, 530-535
     (1974).

10. C. F. Carter, "Trends in machine tool development and
     application," in *Proceedings of 2nd Int. Conf. on Product Development and Manufacturing Technology*, Strathclyde (1971), MacDonald, London, pp. 125-141 (1972).

11. Anon., "NC moves into medium batch production," *Metalworking Production*, Feb. 7, 45-48 (1968).

12. Anon., "DNC for flexibility," *Production*, September, 70-77
     (1974).

13. Anon., "Automation of small batch production," part 2,
     "Technology," *Report of the ASP Committee to the MEMTRB*.
     Dept. of Industry and National Engineering Laboratory, UK
     (1977).

14. J. Hatvany (ed.), *World Survey of CAM*, Butterworths,
     Sevenoaks (1983).

15. J. Browne and K. Rathmill, "The use of simulation modelling as a design tool for FMS," in *Proceedings of 2nd Int.
     Conf. on FMS*, London, IFS (Bedford), pp. 197-214 (1983).

16. Anon., "Kanban — The production control system that
     makes Toyota cars just in time," *Production Engineer*,
     April, 49-50 (1981).

4

Pallets, Fixtures, and Machines

4.1 INTRODUCTION

In this and the next two chapters, the elements of FMS are presented and discussed. The topics covered include the machines,
work-handling systems, pallets, approaches to fixturing workpieces, the layout of systems, and how the systems are manned,
managed, and controlled. These elements are all interrelated,
and there is no inherently logical sequence to their presentation.
However, having so far presented FMS from a top-down standpoint to give an overall perspective, it is now appropriate to
switch to a bottom-up description of the elements and start with
the workpiece because it is the form of the workpiece which de-
termines the type of fixtures, pallets, and machines that are
used.

The two main classifications of workpieces have already been
mentioned. There are turned parts, which have rotational sym-
metry about an axis, and the remainder are referred to as prismatic parts. There is, by definition, an almost infinite variety
of prismatic parts, involving great diversity of size and shape.
Whereas turned parts are mostly produced from bars or sawn-up

bars, prismatic parts may come in the form of castings or forgings
or they may be machined from solid stock. In FMS, a prismatic
part is generally held in a fixture that will take into account the
shape of the part. The fixture is, in turn, mounted on a pallet
which generally becomes the machine worktable when on a machine.

Systems can generally be classified into those which machine
prismatic parts and those which machine turned parts, and it will
be seen that the elements of the two types of system are significantly different when the parts being machined are small or medium size. Because of the differences, the discussion of pallets,
fixtures, and machines will be split between prismatic parts and
turned parts. The elements of systems for machining larger
parts will be considered at the same time, although the differences between the elements of such systems are not typically as
significant.

4.2 PALLETS FOR PRISMATIC PARTS

Various examples of pallets have already been illustrated in the
text when the pioneering FMS were discussed. Pallets form the
interface between the workpiece and the machines, and it is thus
necessary that all the machines in a system accept the same form
of pallet. This is not a problem if all the machines are identical
machining centers. If, however, the system has different machines, then some of the machines must be adapted to suit a
standardized interface.

The pallet configuration generally adopted is that the pallet
has the shape of the worktable of the machining centers of the
system, although there are exceptions to this. Pallets are thus
generally square with large "chamfered" corners and with tee
slots and tenons for positioning and clamping the fixtures. They
usually also have a substantial cross-section, and thus even the
smaller ones can weigh over 100 kg. A typical square pallet,
together with fixture and workpiece, is illustrated in Fig. 3.7.
The underside of a pallet is provided with surfaces so that it
can slide into position on a machine. There, it will generally be
located on conical or tapered locators before being positively
clamped. An air blast is often used prior to clamping to ensure
that all the location surfaces are clean. Although the square
pallet is the most common, larger machining centers (that is,
those over 800-mm^3 capacity) often have rectangular worktables

because larger parts tend to have a major dimension in one direction. Japan and West Germany have standards for square and rectangular pallets which are identical in many details [1]. These look toward the time when greater integration occurs within factories and pallets may need to be used across systems.

Figure 4.1 shows two large rectangular pallets together with parts and fixtures. These are part of the Anderson Strathclyde (Motherwell, Scotland) system supplied during the early 1980s by Giddings and Lewis-Fraser [2]. The system has five identical 150-mm (6-in.) diameter horizontal spindle machining centers and an NC contour boring and facing head machining center. The pallet size is 1200 mm × 1800 mm (48 in. × 72 in.), and the system is used to machine steel castings for a range of long-wall coal-cutting machines. The layout of the system is shown diagrammatically in Fig. 4.2 with the machines installed in a line and supplied with pallets by an automatic guided vehicle (AGV), which is partly visible in Fig. 4.1.

Fig. 4.1 Pallets, parts, and workpiece handling on the Anderson Strathclyde system. (Courtesy Giddings & Lewis-Fraser.)

Fig. 4.2 The layout of the Anderson Strathclyde system. (Courtesy Giddings & Lewis-Fraser.)

A Giddings and Lewis system supplied in 1979 to the Caterpillar Tractor plant in Aurora, Illinois, has even larger pallets (4267 mm × 1524 mm) (14 ft × 5 ft) [2], and in Europe, Caterpillar Tractor uses pallets of 3500 mm × 2500 mm (11.5 ft × 8.2 ft) on a four-machine FMS at their Gosselies, Belgium, plant, the system being supplied by Schmarmann GmbH in 1980. The Aurora pallets have a 20-ton capacity, and this is required to carry substantial fixtures which, in turn, support the frames of large earth-moving vehicles. To avoid having to handle the weight through some other type of intermediate work-handling device, the pallets have wheels and move around a rectangular rail track that is mounted directly on the floor. To visit a machine, the pallets move down a tee leg from the main track and are clamped into position at the machine while still on the track. The workpiece is then machined by traveling column machining centers.

Wheeled pallets are also used in some smaller systems, and this arrangement is also fairly common with transfer lines. A wheeled transfer line pallet has already been illustrated in Fig. 2.2, and the bottom of the same pallet is shown in Fig. 4.3. Its wheels are clearly visible and they are mounted in rotary swivels so that the pallet can move around a rectangular track without the use of turntables at the corners. Hardened steel inserts are visible in a number of positions. These are for locating and clamping the pallet at each workstation and are required to ensure that the pallet has a high resistance to wear on its locating surfaces so that location accuracy and hence machining accuracy are maintained over a long period. Similar features are required on most pallets. The inserts, the machining required for the inserts, the substantial amount of material required to ensure pallets are rigid and stable — all contribute to pallets being expensive.

Fig. 4.3 The bottom of a wheeled pallet. (Courtesy Kearney and Trecker Marwin.)

A few systems use circular pallets. A typical example is il-lustrated in Fig. 4.4, which shows part of a White-Sundstrand system. White-Sundstrand uses circular worktables in its Omni-mill machining centers, which are included in the system illus-trated. This follows the approach already mentioned of giving pallets the same geometry as the machining-center worktables. Circular pallets are also likely to be used in systems that in-corporate vertical turret lathes (VTL), irrespective of the other types of machine in the system. In these cases, the pallets will have the same shape as the VTL worktable.

Figure 4.5 illustrates a two-machine flexible cell, the machines being a vertical turret lathe and a vertical machining center. The cell, engineered by TI Matrix Machine Tools Ltd., is to ma-chine large rotary seals for marine applications [3]. Some of the large rotary pallets, with and without workpieces, are clearly visible. This cell is one of the few examples where workpieces are clamped directly to pallets, effectively dispensing with fix-tures. It is rarely practicable to dispense with a fixture for horizontal spindle machining because the workpiece has to be

Fig. 4.4 The load/unload and pallet storage area of a rail-cart-based FMS. (Courtesy White-Sundstrand, WCI.)

supported in front of the spindle and few workpieces have a geometry suitable for clamping directly to a pallet. It can, however, be practicable with vertical spindle machining, and this includes vertical turret lathes. Saving the cost of a fixture has much to commend it since even smaller fixtures can cost over $1000, and larger ones can cost many $1000s. The combination of vertical turret lathes and machining centers which is present both in this cell and in a number of other larger systems will be discussed later in this chapter.

4.3 FIXTURES FOR PRISMATIC PARTS

It has already been established that most FMS incorporate machines with horizontal spindles and horizontal worktables (the

Fig. 4.5 A flexible manufacturing cell for machining large marine seals. (Courtesy TI Matrix Machine Tools Ltd.)

worktable being the position for the pallet). To hold a workpiece so that it can be machined thus generally involves the use of a substantial, right-angled fixture. The fixture illustrated in Fig. 3.7 is typical of the "picture-frame" type, which is used for locating medium-size castings or forgings that have significant dimensions in all directions. The workpiece fits through the fixture, allowing machining to be readily carried out on the front and back of the workpiece by indexing the pallet. Machining of features on the sides of the workpiece is also often practicable.

A second commonly found form of right-angled fixture can be termed the "pillar" fixture. One form of this is illustrated in Fig. 4.6. The pillar fixture has workpieces attached to its sides and can accommodate two or more workpieces at once, depending on the sizes of the workpieces relative to the size of the pillar. The amount of machining that can take place on any workpiece

Fig. 4.6 A two-workpiece pillar fixture. (Courtesy Fritz Werner, TI Rockwell.)

is only limited by cutter accessibility. For the setup shown, for example, three sides of the workpiece can have boring and similar operations carried out and five sides can be milled. These types of fixture are generally kept permanently attached to their pallets to maintain consistency of part location from part to part.

As workpieces become smaller, pillar fixtures with square cross-sections are generally used and workpieces are attached to all four faces and sometimes also to the top. Figure 4.7 shows a typical example. This is a modular fixture that is configured to hold four workpieces per face, giving the potential of mounting 16 workpieces at a time. In the figure, only two workpieces are shown held in their mounting units. The modularity of this fixture will be referred to again in the next section.

Fig. 4.7 A tooled-up fixture for holding 16 workpieces with pallet and rail trolley. (Courtesy Lucas Group Services Ltd.)

There are significant economic advantages to mounting a number of workpieces on a single fixture. The first of these concerns the basis of all manufacturing efficiency – ensuring high
machine utilization. To have a number of workpieces on a fixture minimizes the time it takes for a machine to switch between
workpieces. It simply requires movement of the machine axes at
rapid traverse or a pallet index through 90°. These movements
take less time than the time it would take to change a pallet.
Some users further reduce dead-time on multiworkpiece fixtures
by minimizing the number of tool changes. They do this by
having a particular tool carry out all the cutting required on the
workpieces on one or all fixture faces before changing the tool.
This requires modularly coded NC programs, but this is quite
practicable with modern computer aids to programming.

The use of multiworkpiece fixtures has a further benefit in
that it lengthens the machining cycle time per fixture. This reduces the number of fixtures and pallets needed by a system with
considerable cost savings. The amount of fixture handling and
the demands on the work-handling system are also reduced, and
this can give savings in both capital and running costs of the
work-handling system. Thus, multiworkpieces fixtures should
be used whenever possible. The main problem associated with
their use is the increased complexity of programming the machining operations.

Although a number of smaller workpieces can be attached to
a single fixture, larger workpieces are mounted one per fixture.
If the workpiece is both box shaped and has adequate strength
to resist machining forces wherever these occur, then its fixturing is simplified and individual elements such as supports,
pillars, and clamps can be used. The Anderson Strathclyde fixture and workpiece illustrated in Fig. 4.1 is an example of such
an arrangement.

4.3.1 Adaptable Fixtures

All fixtures are expensive, and in some ways they can be considered expensive extras. If workpieces can be clamped directly
to pallets, then considerable sums of money can be saved. Because this is not often practicable, it is important that the number of fixtures required by a system be minimized both by the
use of multiworkpiece fixtures and by careful design.

The minimum number of fixtures (and pallets) required by a
system is that number necessary to ensure that the machines

are never kept waiting for workpieces. Thus, there must always
be a loaded pallet waiting at a machine on completion of a machining cycle. The number of fixtures and pallets required will
depend on many factors, including the cycle times of the workpieces, whether the workpieces go to the next machine in the
same fixture, the load/unload cycle times, the type and speed
of the work-handling system, the number of machines in the system, and the system schedule.

The schedule of the system must take into account which workpieces can be held in which fixtures. Thus, fixtures are a critical resource and they can directly limit the rate of production of
any part. If a system produces a small range of parts (say nine
or less) and if the part schedule varies only slightly, then dedicated fixtures are probably acceptable with two or three fixtures
(and pallets) being provided for each part. The range of parts
is produced all the time and must be produced all the time to keep
the machines busy. However, if a change of schedule means that
more of one part type is required, then the dedication of the fixtures immediately limits the ability of the system to respond to
the new schedule. One approach to solving this difficulty is to
provide spare fixtures and pallets to enable a system to respond
to a varying schedule. Most of the pioneering FMS were designed
to machine a restricted range of parts, and the part mix did stay
reasonably constant. The provision of one extra fixture for a
particular part enabled that part to be produced in considerably
higher quantities, and with a small number of part types, the
cost of extra fixtures and pallets was not excessive. As FMS
started being used for larger ranges of parts, this solution became increasingly costly, and some means had to be found of
having adaptable fixtures that would accept more than one type
of workpiece.

Various approaches have been taken to designing adaptable
fixtures. Many of these have been applied to the fixturing of
smaller parts and are based on pillar fixtures. The fixture in
Fig. 4.6 can be seen to have a matrix of predrilled and prebored
holes which are used to locate and secure adaptors for particular
workpieces. The adaptors only have to be changed when the
part mix changes, and an adaptor plate is obviously comparatively
inexpensive compared with a whole fixture. This type of approach is also used on the Vought system of Dallas [4]. For this
FMS (illustrated in Fig. 1.1), seven types of pillar fixture were
initially designed, which between them enabled 500 different workpieces to be machined. As the unmachined parts were mostly

billets of aluminum alloy, the fixturing task using the predrilled
holes was not as demanding as some. If does, however, illustrate what can be achieved.

Figure 4.7 shows another example of a fixture designed with
adaptability in mind. Close examination of the fixture will reveal
that the elements of the fixture have been assembled onto the
sides of a pillar cube. The parts are held in mounting units
which are individually attached to plates, and the plates are, in
turn, clamped to the sides of the pillar cube. (Apart from the
base, the pillar cube itself is totally obscured by the attachments.) Various elements of this fixture could thus be disas-
sembled and reused in a different fixture. Any of the mounting
units or the plates could also be changed to suit other part(s).

The configuration of a fixture is only changed, however, when
absolutely necessary. It not only takes time to reconfigure a fixture, but rebuilding a fixture can result in small positional variations that alter the location of parts with respect to the pallet and
hence to the machine spindle, which can result in machining errors. To overcome this difficulty, it is usual practice to gauge
exactly where each face of a reconfigured fixture is with respect
to the spindle by use of a spindle-mounted probe. The appropriate offsets are then used automatically during machining to
compensate for the errors.

The best form of adaptable fixture is one designed to accept
a variety of parts without requiring reconfiguration. This approach generally has to be used with larger parts, castings, and
forgings, which pose more of a problem. It is a problem that has
to be addressed in both the design and manufacturing offices if
adaptability is to be achieved. It can be achieved if both the
fixtures are designed to fit the workpieces and the workpiece designs are adapted to suit the fixtures. With systems often being
used to machine a family of parts, this is often possible with suitable collaboration at the early stages of workpiece design.

4.4 PALLETS AND FIXTURES FOR TURNED PARTS

Small and medium-sized turned parts are machined on lathes with
the lathe providing the fixture. (Subsequent operations may involve other types of machine.) The part fixture will generally be
a chuck or a combination of chuck and tailstock center, and these
are parts or accessories of the machine rather than being workpiece related. The only time a turned part needs anything extra

attached to it before it is mounted in a machine is when it is held between centers and a driving dog is needed. To handle such a part when loading and unloading a machine, an industrial robot will generally be used and it will grip the part directly.

Figure 4.8 shows a two-machine cell incorporating a Cincinnati T³ industrial robot. The cell can act as a flexible manufacturing cell (FMC) or as a batch manufacturing cell with parts having their first operation on one machine and the second on the other. The workpieces arrive on a conveyor on tray-type pallets (visible in the foreground). The robot is programmed to pick up each part in turn from its known position on the pallet tray at the end of the conveyor. A limit switch is generally used to

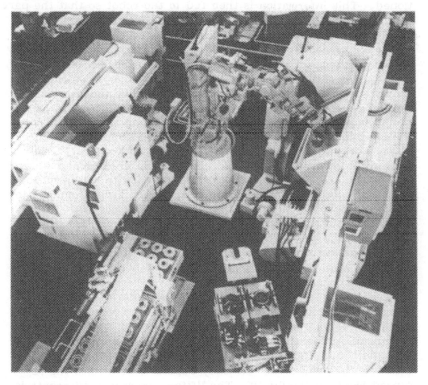

Fig. 4.8 A two-lathe robot-served cell. (Courtesy Cincinnati Milacron.)

indicate whether a pallet is present or not. In a flexible cell, a
bar code on the pallet might indicate to a central computer which
pallet had arrived. The computer has data on the contents and
layout of the pallet already stored so that the appropriate sets of
programs can be passed both to the robot to unload the pallet and
to the machines to machine them.

Some robots now have vision systems attached that enable them
to determine the type, position, and orientation of parts. A tele-
vision camera is used to view the parts on a pallet or on a conveyor. The camera generates an image of the parts in its field
of view, and these are analyzed in turn by image analysis software to determine which part it is, where it is, and its orientation. Images of the parts in various possible orientations have
to be stored on the robot computer so that the robot and the system can work out which part is being viewed and how it is positioned. This information is then fed to the robot so that the part
can be picked up.

Figure 4.9 shows another pallet arrangement, one used by
Mazak for turned workpieces [5]. Again, the workpieces occupy
known positions on a pallet, which, in this example, is circular.
In both this and the previous example, the workpieces can be
simply supported at known positions on a pallet and thus incur
minimal pallet costs. Furthermore, a very basic means of location can be used to position the workpieces on the pallet. For
the two types shown, simple tapered spigots to locate the bores
of the workpieces are all that is required. In contrast, small
shafts will generally be stacked vertically in crate-type pallets,
and plastic crates or wire baskets can be used to hold a number
of shafts. The form of the crate can be very simple, and its top
need only have a board or piece of sheet metal across it with
holes to locate the tops of shafts. A similar simple means of location must be provided to locate the bottoms of the shafts. As
shafts increase in size, so tray-type pallets are again likely to
be used, with the shafts being supported horizontally beside
each other on horizontal vee supports.

Multiaxis industrial robots are very suitable devices for handling small and medium-sized turned parts, as robots have the
flexibility to pick up a wide variety of parts in a variety of orientations. Different grippers may be needed to handle some of
the parts, but gripper changing from a gripper store beside the
robot can be programmed on many robots to give them the required handling flexibility. The lifting capacity of all robots is
limited, but with turned parts needing no additional fixturing

Fig. 4.9 Circular pallets with turned parts. (Courtesy Yamazaki Machine UK Ltd.)

and generally weighing less than 75 kg, most turned parts can be handled.

Flexible chucking on a lathe for different diameter workpieces poses a problem because most automatic chucks only have limited travel on their jaws. Various solutions to this problem are available but none is yet particularly flexible, or quick. As chuck jaws are stepped, it is possible to use one set of jaws for two or three parts of different diameters. This provides only limited flexibility and is only a partial solution. One approach to a complete solution to the problem involves automatically changing complete chucks so that the right size jaws for a different part come with another chuck. Arrangements also exist for automatically

changing each jaw in turn (rather than a complete chuck) and
for automatically fitting inserts into more permanently attached
jaws. Flexible fixturing is thus a problem that has been adequately solved for lathes, but the solutions are not particularly
elegant. Changing chucks or chuck jaws takes time and the time
is nonproductive. It must be remembered that small and medium-
size turned parts can be produced quickly on modern lathes. The
time it takes to change workpieces and attachments as a proportion of the part cycle time thus becomes increasingly significant
as cycle time reduces.

Without any switch of part type, the time between parts is
mostly determined by the load/unload robot. The quickest way
of changing workpieces using a robot is for the robot to have a
double indexing gripper. One gripper is used to receive the finished part while the adjacent gripper is preloaded with the next
part. The movement of the robot required to change workpieces,
and thus the time required, are minimized with this arrangement.
To apply this method to changing a range of parts in any sequence would require having double grippers for all combinations
of part types. This would obviously be very expensive as many
more grippers would be required in the various combinations necessary. The alternative of changing single grippers between unloading one part and loading the next is far less costly but is
very time consuming.

Thus, although both robot grippers and lathe chucks can be
designed to be flexible, the time it can take to achieve the flexibility means it is generally sensible and economical to batch
turned workpieces. If chucks are to be changed to accept different workpieces, then it is arguably better to preload the
workpiece into a chuck prior to loading and avoid all the robot
gripper problems by simply addressing the problem of handling
the chucks. This approach resembles that of using rotary pallets for machining prismatic parts on vertical spindle lathes.
One FMS user is known to be approaching this by using an interchangeable face plate as a pallet on a horizontal spindle lathe.

4.5 PRISMATIC PART MACHINES

The systems already discussed have contained a number of different machines, with the size and type of machine naturally reflecting the form of the workpieces to be machined. As stated
previously, there are far more FMS for machining prismatic parts

than for machining turned parts, so the machines for prismatic systems will be considered first. The workpieces are generally either castings or forgings, although some aerospace applications involve machining parts from solid. Systems exist for machining parts that would fit into a 150-mm cube up to systems of the capacity supplied to Anderson Strathclyde and Caterpillar Tractor, which were considered earlier in this chapter.

Most systems are planned to machine a family of parts so that workpiece size, operations required, material, and geometrical tolerances do not vary too greatly. Although many small workpieces are produced on larger machines, there is a limit to the part size variation that can be accommodated on a system. There comes a point when the definition of a "family of parts" has to distinguish between workpieces on the basis of size. This is partly a matter of machine economics and partly a technical problem of tooling. The use of a number of large, high-powered machines to machine large and small workpieces is economically wasteful when the smaller workpieces are being machined. A more significant problem arises, however, in the variety of tooling likely to be needed; this is discussed in the next section. Although it is discussed with reference to machining centers, the comments also apply to other machine types.

4.5.1 Machining Centers

It has already been stated that the flexibility of horizontal machining centers (HMCs) makes them the foundation of FMS, and many FMS have only these machines. Vertical machining centers (VMCs) are rarely used. They have the advantage of requiring only simple fixtures, but the disadvantage of the spindle not having the easy access to a number of sides of a workpiece at a single setup, as is possible with an indexing worktable on an HMC.

Figure 4.10 illustrates a typical modern HMC with two tool chains and a pallet storage carousel. Whereas the HMCs of the 1960s might have had only a 20-tool carousel, some modern HMCs used in FMS have provision for over 100 tools. There are two main reasons for providing the increased tooling capacity. The first of these relates directly to flexibility. The more tools available on a machine, the more workpiece types it should be capable of machining and the greater the flexibility of the machine and the system. The second reason relates to unmanned system opation, which users are increasingly planning for. It must be

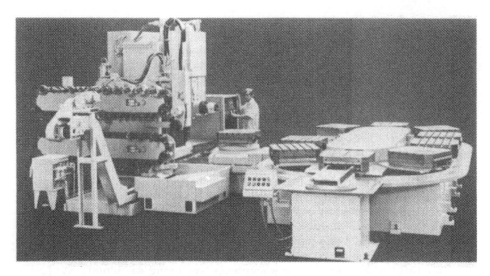

Fig. 4.10 A horizontal machining center with pallet storage carousel. (Courtesy Cincinnati Milacron.)

realized that up to the 1980s, the system tooling was typically manually supervised by system operators, and tools were re-placed as necessary by these operators. If machines are to op-erate unattended for long periods, then all the tools needed for that period must be prestored at the machine and this requires the necessary storage capacity. Some tools may require dupli-cates (or "sister" tools) if their life is shorter than the unat-tended period, and this poses additional demands.

Just as a system should be designed with a minimum number of machines to meet a given production requirement, so should the number of tools required be minimized. However, the demand for greater flexibility and for periods of unmanned operation has increased the tooling generally needed. These conflicting de-mands are posing problems for a number of system users, and the conflict is often being resolved by specifying extra tooling and larger carousels. This adds to the capital cost of a system but it appears to be a price users are prepared to pay. Systems are gradually being implemented that will automatically swap tools in and out of a carousel and thereby reduce the need for large tool storage at machines, but this development is progressing slowly. This and other aspects of tool handling and the management of tooling are covered in Chapter 6.

The pallet storage carousel shown in Fig. 4.10 is designed to exchange pallets with the machine, provide storage for unmachined workpieces, and finish machined workpieces. When fully loaded with unmachined workpieces, it enables a machine to operate unattended for a considerable period. The incorporation of storage carousels into systems is considered further in the next chapter.

There are workpieces and production requirements that lead to machines other than machining centers being included in FMS. Their inclusion is generally a reflection of limitations in machining centers because it is a basic premise of FMS design to have as many machines identical as possible. These limitations usually take the form of productivity limitations, accuracy limitations, or machining limitations. Other machines found in FMS include multispindle head indexers, head changers, and vertical turret lathes, and these are now considered.

4.5.2 Head Indexers and Head Changers

Multispindle head indexers are capable of drilling, tapping, and boring using any of a number of preset heads arranged on a large indexable turret. A typical machine is shown in Fig. 4.11. The heads resemble unit heads of transfer machines in having a number of fixed spindles that can produce a number of holes or bores simultaneously. If a pattern of holes is to be produced, then this will generally be done with the ends of the drills supported by a drill-bush plate that abuts against the workpiece. Head indexers enable a large number of various size holes to be produced in a single drilling cycle with the accuracy and repeatability inherent in the use of drill-bushes. When this is compared with drilling on a machining center, it will be apparent that the cycle time will be approximately N times shorter using a head, where N is the number of drills involved. Furthermore, the location accuracy of any hole and its relative position to any other hole will be superior using the drill-bush plate because every drill is constrained at the workpiece.

Boring operations can be similarly improved, not by using bushes but by incorporating robust, balanced-cut boring tools into a head. Once set up, these can more easily achieve tight tolerances. The setting and design of heads permit the maintenance of closer bore-to-bore tolerances than is possible with

Fig. 4.11 A head indexer. (Courtesy Kearney and Trecker
Corp.)

most machining centers, particularly when operating in an FMS.
Head indexers usually have only two degrees of freedom; one
about the indexing axis and the other the linear movement of
the worktable toward the indexing axis. The workpiece on its
pallet is brought to the indexer workstation and then fed toward
a head for machining to take place.

Head indexers are built into an FMS when the system has to
machine a limited range of parts that require hole patterns and
precision boring. The limited range of parts means that each is
produced in large quantities, and it is worthwhile being specially
tooled-up to produce hole patterns that have to be machined very
frequently. The alternative of drilling the holes one by one on
a machining center is not a very efficient means of utilizing a
machining center and, because of the time it would take, would

necessitate an increase in the number of machining centers in a system.

The Allis Chalmers system for machining transmission parts for tractors (described in Chapter 3), the Avco Lycoming system (at Williamsport, PA) for machining aircraft engine parts [6] and the John Deere Component Works system (at Waterloo, IA) for machining tractor parts are examples of systems with head indexers. The Allis Chalmers system shown in Fig. 3.6 includes eight head indexers arranged in duplex pairs with shared workstations. The pallet can be indexed on a rotary table at the workstations so that the appropriate side of the workpiece faces the appropriate head. The pallet can be fed in both directions so that either indexer can machine the part.

Head changers are similar to head indexers in principle but different in design. Rather than heads being fixed in a turret configuration, individual heads are brought from a storage system and attached to a drive unit at the single workstation. The number of heads that can be used is thus limited only by the head storage capacity. Figure 4.12 illustrates part of the storage area of a head-changing machine. It is of somewhat untypical design, as the heads are stored on their backs. Many designs store the heads in their cutting orientation. Head changers have been developed by a number of companies primarily as a means of introducing flexibility into transfer lines. Their application to high-volume production in FMS followed from this, and they have also found application as stand-alone machines [7]. The productive advantage of unit heads has been recognized by some manufacturers of large machining centers who have designed their machines so that they can accept and automatically change small unit heads.

Machining tolerances with head changers are almost as good as with head indexers, and production rates are similar. Thus, both machines offer accuracy and productivity advantages over machining centers. Head changers and head indexers are not, however, as flexible as machining centers. Product changes or changes in market preferences can suddenly make a set of unit heads obsolete, and the productive capacity of an FMS can suddenly be reduced as a result. A new head for either type of machine is considerably more expensive to make and takes longer to produce than the time it takes to acquire new tools for a machining center and write a machining program. New heads also effectively freeze a design. They thus have a disadvantage that FMS are intended to avoid. The configuration of heads should therefore

Fig. 4.12 Part of a head changer used for storing heads. (Courtesy Buckhardt and Weber, Hahn & Kolb.)

only be finalized once a product has passed through its development and market acceptance stage, when further design changes are unlikely.

4.5.3 Vertical Turret Lathes

Vertical turret lathes (VTLs) can be considered hybrid machines in that they are often used for turning prismatic parts. They are restricted, however, to turning larger prismatic parts with bores that cannot readily be bored on a machining center. Products such as valve bodies, transmission casings, and covers often have large bores and adjacent annular surfaces which are easily and quickly produced by boring and facing on a vertical lathe.

The Matrix system illustrated in Fig. 4.5 includes a VTL, and the workpiece form is obviously ideally suited for being produced

by turning. The earliest FMS purchased by Caterpillar Tractor
from White-Sundstrand (ca. 1974) included two VTLs along with
four Omnimill HMCs and three Omnidrills [8]. The VTLs were
for carrying out the large boring/turning operations, and the
drills were included to relieve the machining centers of simple
drilling tasks. Ingersoll has supplied small systems (in terms
of the number of machines) which have included VTLs, but considering the number of FMS involved in machining transmission
parts, there are fewer VTLs in systems than might be expected.
The reasons for this relate to two difficulties with the use of
VTLs and to two fundamental principles of FMS design. The
difficulties will be considered first.

The first difficulty relates to workpiece geometry. Although
some valve bodies and transmission housings have centrally positioned bores, other workpieces may have them off center, sometimes substantially. To mount such workpieces on the bore axis
on a VTL creates a very unbalanced situation. This can be cured
by providing a dynamically equivalent balancing mass, but this is
expensive. An alternative approach is to keep cutting speeds low,
but this is inefficient. A workpiece with an off-center bore can
also overhang the pallet and cause handling problems.

Another difficulty with a VTL is that the turning axis is fixed
for each fixture setup. Thus, housings with more than one bore
axis require different setups and probably different fixtures for
each axis. This results in more handling, additional load and unload effort, and additional fixtures, all of which add to the cost
of manufacture.

Large bores and similar workpiece surfaces are increasingly
being produced by circumferential milling on machining centers.
This method may take longer than by using a VTL, but it avoids
the difficulties just discussed and it helps to minimize machine
variety — an important design principle of FMS. A related design principle is to incorporate redundancy into a system so as
to ensure a system only "gradually degrades" on failure of any
of its elements. The provision of redundancy when considering
whole machines means ensuring that another machine can carry
out the tasks of any machine that is temporarily out of commission so that production can continue, even if at a reduced level.
Some systems satisfy this principle by including a minimum of two
machines of any one type within a system. A system obviously
has most redundancy and maximum flexibility if all the machines
in a system can deputize for any other.

Two problems of VTLs have already been discussed and a
third can be considered to be that they are not a machining

center. This objection could, of course, be made to most other types of machine, and it would perhaps be better to restate the objection as: they lack flexibility. An HMC can deputize, if necessary, for a VTL, but a VTL can only turn and bore.

4.5.4 Other "Prismatic" Machines

Having just stressed that machine variety should be minimized, users have found reasons to have a number of other machines in systems for machining prismatic parts. Among other machines that may be found are CNC drilling machines, CNC milling machines, and precision boring heads. The CNC drills and CNC mills are effectively specialized machining centers and are likely to be included to save the cost of machining centers. Both machine types will be incorporated to reflect the machining needed on the initial products it is planned to machine. For parts that are robust and can take high rates of milling prior to more delicate finishing operations, a good case can certainly be made for incorporating a higher-powered CNC mill for roughing operations. Milling and drilling machines can usually machine parts on the same fixtures as a system's machining centers, so the use of these machines does not involve extra expense in other areas. Tooling considerations may also encourage their use. However, whether there is a real cost saving in having less versatile machines and compromising the flexibility of a system to accept different parts in the future is something that each customer has to consider.

Precision boring heads are used very occasionally for producing very high-quality bores. They have no flexibility, however, and might better be used outside a flexible system as an independent finishing operation.

Some of the above observations are reflected in the machines incorporated within the GE system in Erie (PA) [9]. This system otherwise appears to be an example of a system that breaks the rule of minimizing machine variety. The $16 million system was supplied by Giddings and Lewis in the early 1980s to machine six families of locomotive motor frames and gearbox parts. Its layout is shown in Fig. 4.13. The system has nine heavy-duty machine tools — three heavy-duty horizontal boring mills with 180-mm (7-in.) diameter spindles and 37.5-kw (50-hp) motors for roughing work, two special vertical milling machines for carrying out precision internal machining, one horizontal-spindle boring mill with a 150-mm (6-in.) spindle for finish boring and

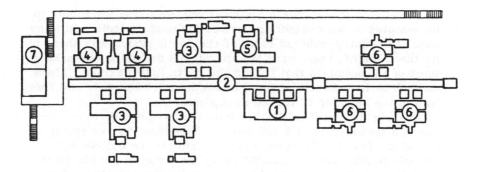

Fig. 4.13 The layout of the GE, Erie, FMS. 1, Load station; 2, rail cart track; 3, horizontal boring mills; 4, vertical milling machines; 5, horizontal boring mill; 6, horizontal machining centers; 7, control room. (Courtesy Giddings & Lewis.)

milling, and three horizontal machining centers with 11-kw (15-hp) motors for drilling, tapping, counterboring, spot facing, and other precision work. The machines can be seen to be split between three roughing and six finishing machines, with four of the finishing machines having horizontal spindles and two being identical vertical-spindle boring mills. This system thus appears to have adequate machine redundancy and to be fully capable of operating with any machine temporarily out of commission. The system work handling will be considered in the next chapter.

4.6 TURNED-PART MACHINES

There are very few FMS for turned parts, although there is an increasing number of two or three machine flexible turning cells served by a robot, as illustrated earlier in this chapter. During the early years of FMS, the Fanuc system described in Section 3.2.3 and an East German "Rota" system were the only examples of turned-part FMS. It is believed that there are three main reasons for this inbalance in the development of the two types of system. First, prismatic parts generally have higher initial costs, higher added values, and longer manufacturing lead times than turned parts. There was thus far more incentive to reduce their lead times than with turned parts. Second, the development of FMS preceded the development of cheap robots suitable for work

handling, and it also took a comparatively long time for robots to
be accepted as work-handling devices for machine tools. In many
ways, the gantry-mounted robot of the Fanuc system was ahead of
its time. Third, there were the operational difficulties of making
many of the machines that cut turned parts fully flexible. These
difficulties have been discussed in Sections 3.3 and 4.4, and they
lead to the batching of turned workpieces.

Bearing these difficulties in mind, the machines that may occur in a turned-part FMS will now be considered by reviewing
one of the few FMS for turned parts. This is the British 600
Group's SCAMP system, SCAMP being the acronym for Six (Hun-
dred Group) Computer Aided Manufacturing Project [10]. The
"600 Group" is a conglomerate with a substantial machine tool manufacturing and robot marketing business.

4.6.1 The SCAMP System

The SCAMP system is shown in Fig. 4.14 and consists of nine machine tools that are robot loaded with parts brought to the machines in pallets carried by a computer-controlled carry-and-free
conveyor. Four conveyor tracks can be seen in the figure. The
nine machines are arranged in a line parallel to and to the left of
the conveyor tracks, as viewed in Fig. 4.14. The two central
conveyor tracks are linked at each end and form a continuous loop
for recirculating the pallets. The outer tracks have gates to the
inner loop and provide buffer storage. The left-hand track provides pallet storage and buffers for the nine machines, the right-
hand track for six load/unload stations. The conveyor can accommodate 150 pallets, each of which is individually coded and
under the control of the supervisory computer. Tray- and crate-
type pallets for disk-shaped and shaft workpieces, respectively,
can be seen on the conveyor. Workpieces may be of steel, cast
iron, or aluminum and are machined in batches varying from 25
to 100. Nine different batches may be in the system at a time.
The system is usually manned by three operators.

The system was designed to produce a wide variety of turned
parts. It has the following machines: two two-axis Colchester
650 CNC lathes, two five-axis Colchester 650s CNC lathes, a Sykes
CNC gear-chamfering machine, a Sykes Genertron gear-shaping
machine, a Matrix CNC cylindrical grinding machine, a Sykes
H160 hobbing machine, a Clarkson broaching machine. The five-
axis lathes are modified two-axis machines to allow them to do
some milling and drilling. The gear shaper is partly numerically

Fig. 4.14 A view of the SCAMP system. (Courtesy *Metalworking Production*.)

controlled in that it can receive positioning data from the supervisory computer. However, the hobber and broaching machines both must be set up manually for the particular workpieces they are machining. There are thus limits on system flexibility.

Because turned parts are handled individually and because of the flexibility of the robot loaders, this range of machines can be readily incorporated into the system. It does not matter that the machines hold the parts in a number of different configurations and at different heights. This contrasts directly with most prismatic part FMS, which require a standard interface for all the machines in the system and a common worktable height.

The use of robots thus readily facilitates almost any type of machine being incorporated into a turned-part system. This being so, individual machines that may form part of a turned-part FMS will not be considered further. The crucial questions with any machine tool under consideration for a turned-part FMS are: Is its flexibility limited in terms of its control, its fixturing, or its tooling? And if it is, do these limitations restrict the specification and operation of the remainder of the system?

This completes the review of metal-cutting machine tools commonly found in FMS. Another type of machine increasingly being found in FMS is a coordinate measuring machine. This will be considered in Chapter 6 together with systems that include other types of process. Chapter 5 reviews how systems are laid out, the variety of system forms, and means of work handling.

REFERENCES

1. R. Lewald, "Pallet standard catches on in FRG," *American Machinist*, November: 104—105 (1985).

2. D. J. McBean, "Practical applications of FMS," in *Proceedings of 2nd Int. Conference on FMS*, IFS (Publications), Bedford, London, 477—484 (1983).

3. Anon., "FMS for underwater seals," *Chartered Mechanical Engineer*, May: 32 (1985).

4. Anon., "Today's FMS: Vought Corp.," *Iron Age*, Aug. 16, 33 (1985).

5. Anon., "Mazak's FMS: Making machine tool manufacturing look easy," *Manufacturing Engineering*, September, 63 (1983).

6. Anon., "Our 'FMS' will do the work of 67 conventional machine tools," *Production*, April: 66—69 (1978).

7. Anon., "More flexible automation with head changing machines," *Manufacturing Engineering*, December: 51—52 (1977).

8. Anon., "DNC for flexibility," *Production*, September: 70—77 (1974).

9. Anon., "Today's FMS, General Electric-Erie," *Iron Age*, Aug. 16, 36 (1985).

10. Anon., "SCAMP is revealed," *Production Engineer*, January 14—16 (1983).

5

Work-Handling and System Layouts

5.1 WORK-HANDLING EQUIPMENT

System integration is achieved by two means: mechanically by
work-handling equipment and electronically by computers and
data transmission links. The topics of this chapter all have
some relation to work-handling equipment; aspects of computer
integration are covered in the next chapter. The primary purpose of work-handling equipment is to transport pallets and
workpieces between the load/unload stations and the system machines, and this was its only role in most of the early systems.
As will be explained, the role of work handling has gradually
increased as designers have expanded the ideas behind flexible
manufacturing. For example, work-handling equipment is also
being used to handle tooling and to integrate various forms of
storage into systems as well as sometimes providing the storage
directly itself.

Most of the work-handling equipment used in FMS has already
been mentioned in systems described in earlier chapters. This
has included conveyors, tow and rail carts, guided vehicles,
pick-and-place units, and robots. In the following paragraphs,

this equipment will be reviewed in more detail and the advantages of using certain types of equipment for particular applications will be given. Because the type of work-handling equipment used has a significant influence on how a system is physically laid out in a machine shop, the layout implications of using a particular type of equipment will be covered at the same time.

Although most existing FMS have been built as complete systems, a number have been built in phases, and there are considerable attractions in this approach, particularly for smaller companies. Smaller companies cannot always finance a large-scale system at one time but want to gain experience of the technology and progressively move toward flexible manufacturing. A modular means of implementing FMS is very important for such companies, and types of work-handling equipment and layouts are increasingly being selected with this in mind. Modular design is the final topic covered in this chapter.

Before specific reasons for selecting a particular work-handling system are discussed, a significant influence on the selection should be pointed out. This can be considered a principle of "safe" design practice, and it applies to any form of innovative design. It can be stated as follows: When advancing the state-of-the-art, do not use new technology just because it is new technology or try to advance on too many fronts simultaneously if this is not necessary.

Thus, in designing a novel system, it is wise to incorporate as much proven technology as possible to ensure that development effort can be concentrated on the really new features of the system. Proven technology is sometimes referred to as "retained engineering." Thus, when a company is considering the type of work handling to use, the reason it selects a particular type of equipment may be because it has used it before and is confident it can use it again easily and successfully. This is not a technical reason in the normal meaning of the words, but it is a sound technical reason as far as a particular supplier is concerned. It is in both the suppliers' and the customers' interest to use proven technology because the development and commissioning phases of implementing a system are shortened.

In this and other chapters, generic names (such as rail cart) have been used in referring to particular types of work-handling equipment. Suppliers of such equipment sometimes use more elaborate names to create a high-tech image. These names can be misleading if the form of the equipment is not fully described.

For example, a rail cart may be referred to as a "robot cart" although it may have none of the features of an industrial robot, except the ability to respond to programmed instructions.

5.2 ROLLER CONVEYORS

Belt and roller conveyors were well established means of work handling long before FMS were thought of. Belt conveyors are generally used for bulk handling, whereas roller conveyors are more commonly used for discrete part handling. Four of the pioneering systems had roller conveyors. These were the Sundstrand Aviation (SA) system (illustrated in Figs. 3.1 and 3.2), the Herbert Ingersoll (HI) system (illustrated in Figs. 3.3 and 3.4), the system supplied by Sundstrand to Ingersoll Rand, and the Borg Warner system (illustrated in Fig. 3.10). The more recently developed SCAMP system (illustrated in Fig. 4.14) also uses roller conveyors, as do other, more recent systems that will be described. Although the use of conveyors as retained engineering may well have been significant in all these systems, another significant factor that made their use practicable may be seen by examining the illustrations of the HI, SA, and SCAMP systems — i.e., these systems have to handle only comparatively light loads, with all three systems using lightweight pallets and handling small and lightweight parts.

Conveyors support their loads well above the floor, and the heavier the unit loads, the more substantial and hence costly does a conveyor and its supports become. Many of the other systems that have been reviewed have large pallets and carry heavy workpieces in amply proportioned fixtures. The cost of conveyor systems for such combinations will generally be high compared with alternative approaches in which the weight is transferred more directly to the floor, generally through some form of vehicle.

Although wheeled vehicles of various types are now the work-handling system most frequently used in FMS, conveyor systems are still relevant and will continue to be used where unit loads are not too high. This is exemplified by their recent use in the SCAMP system and in two other European-designed systems for machining prismatic parts. The first of these was the system supplied to Gardner & Sons of Manchester, England, by Kearney and Trecker Marwin (KTM of Brighton, England) [1]. The second, which followed some 2 years later, was the system supplied to Borg Warner at York (PA) by Comau (of Modena, Italy) [2,3].

5.2.1 The Gardner FMS

The Gardner FMS has many interesting features. A view of the system is shown in Fig. 5.1 looking down the conveyor, which links seven KTM machining centers. The system was designed to machine diesel engine crankcases made of aluminum alloy, and initially three types of crankcases were to be machined. All six faces of the crankcases have to be drilled, tapped, and bored, with the machining required being divided into two operations. The first operation consists of machining the sides and ends of the crankcases; this is performed on one of the first three machining centers. These machines have indexing worktables permitting machining on all four faces. The second operation consists of machining the top and bottom of the crankcases and drilling angled holes in the sides and ends. This takes place on any

Fig. 5.1 A view of the Gardner FMS. (Courtesy Kearney and Trecker Marwin Ltd.)

of the remaining four machines. For this operation, the workpieces are held in trunnion fixtures that rotate the crankcases
about a horizontal axis. There is a five-station buffer store between the first- and second-operation machines.

The crankcases are mounted on what can best be described as
an adaptor-fixture rather than a pallet-fixture. The adaptor-
fixture takes the form of an interface that fits onto the base of
the crankcases and provides location features for the combination to be attached to other fixtures mounted on the machine
worktables. The adaptors are lightweight because they do not
have or need much intrinsic strength (as the crankcases provide this themselves), because they have a thin cross-section,
and because they have a substantial central opening so that machining can be carried out on the bottom of the crankcases.
Thus, although the crankcases are large, the fact that they are
aluminum and are mounted on light adaptor-fixtures rather than
large, heavy pallet-fixtures makes the unit loads acceptable for
a conveyor system.

Unlike the other conveyor-based systems already described,
the Gardner system is laid out in a single straight line, with input at one end of the conveyor and output at the other. Parts
pass down the line once, and the adaptor-fixtures are returned
to the start of the line by an overhead conveyor, mounted above
the roller conveyor. These features and the fact that the initial
production was of just three part types bring the system closer
to being described as a flexible link line rather than an FMS.
The distinction drawn here is that other FMS have provision for
machines to be visited in any order rather than in a sequence.
The use of the machining centers does, of course, provide a
great deal of potential flexibility, even if this is not fully used.

Another noteworthy aspect of this system is that the seven-
machine FMS is part of a larger 14-machine system for completely
processing the crankcases. The crankcases come to the FMS after
preliminary rough machining on a line of four interconnected Bohle
machining centers and operations on a multihead drill and a boring
machine. On leaving the FMS, the crankcases go through a wash,
sample inspection, fine boring, and a further final wash. This
integration avoids what has been referred to as the creation of
"islands of automation." An island of automation is created when
just one part of a process is automated. Although this island may
give lead time reductions and productivity improvements of over
500%, the effectiveness of these improvements is dissipated if
other parts of the total processing are prone to result in delays

and inefficiencies. Herein lies a dilemma for management, which will be discussed at the end of this chapter. The 14 machines of the complete Gardner system replaced 75 conventional machines and resulted in double the production capacity. Twenty-two operations were previously carried out on the crankcases, and a crankcase could be produced in 18 hours. A crankcase can now be produced in 6 hours and work-in-progress has been minimized.

5.2.2 The Borg Warner FMS

The Borg Warner FMS comprises five CNC machines, a workpiece load/unload station, and a three-station washing machine, all linked by a roller conveyor. The roller conveyor is arranged as a rectangular loop with long sides and short ends. The machines are situated along one of the long sides; Fig. 5.2 shows a view of the system down the length of conveyor adjacent to the machines. These are four traveling column machining centers and one vertical turning lathe. Each machining center has two chain-type tool storage carousels containing 140 tools in total. One of the carousels is mounted on the column and contains the most frequently used tools. The second carousel is floor-mounted. The washing stations and load/unload station are served by the other long side of the conveyor. The layout thus has some similarity with the SCAMP system in keeping all the machines to one side of the conveyor loop.

In many ways, this conveyor-based system provides a complete contrast to the Gardner system. The system was designed to machine 85 different workpieces, which are parts of commercial air-conditioning and refrigeration equipment. The parts make up seven "families." Many of the parts are fairly small and can be multifixtured, as seen in the figure, using seven different fixture types. Thirteen pallets on average circulate around the conveyor loop at a time, and up to 22 can be accommodated.

The storage capability of conveyors can be an important feature. In later sections of this chapter, various types of storage device will be considered which are generally stand-alone storage devices. As such, their cost relates to the structure needed to support loaded pallets at an appropriate height for other forms of work-handling equipment to access them. The fact that conveyors can partly provide system storage while also acting as work-handling systems improves the economics of their use. It is interesting to note that Comau has supplied other companies

Fig. 5.2 A view of the Borg Warner FMS. (Courtesy Comau/ RSLA Marketing International Ltd.)

with conveyor-based systems, and thus conveyors are proven technology for them.

The main disadvantage of conveyor systems is that they obstruct the floor area and limit access to the front of machines for maintenance or repair. This is particularly true of conveyors used with machining centers, as the machines invariably are positioned right up to the conveyor. Machines that are robot-loaded (as in the SCAMP system) are not necessarily as obstructed. The main alternative to conveyors is some form of wheeled cart or vehicle. There are a number of forms of these, most of which permit easy access to machines. Wheeled carts only rarely deliver a pallet directly to a machine; rather the pallet is delivered to a machine shuttle, which in turn supplies it to the machine. Before carts and other types of vehicle-based work-handling systems are described, it is important that the functions and types of shuttle used in systems are understood, as they have significant influence on system layout and on how the various types of wheeled work-handling equipment interface to machines.

5.3 MACHINE SHUTTLES AND SYSTEM STORAGE

Machine shuttles are found in most systems that machine prismatic parts mounted on pallets. In their simplest form they consist of pallet changers, as illustrated in Fig. 2.4. They serve the important function of providing input and output buffers to the machines and are thus a form of in-process storage.

The machines are the productive elements in any system, and it is important to maximize their utilization. In most systems, this is done by providing machines with shuttles to accommodate two pallets, one space being kept empty to receive the pallet being machined on completion of its machining cycle, the other to be loaded with the next pallet so that it is immediately available. On completion of the pallet change, the work-handling system is signaled that there is a pallet to be collected and that another workpiece will subsequently be required. With the two-space shuttle, the work-handling system has to remove the machined part and provide a new part within the cycle time of the part being cut. Shuttles are thus a means of decoupling the operation of the machines from the scheduling of the work-handling system, and they thus enable the utilization of the machines and the work-handling system to be maximized separately without one significantly interfering with the other.

The amount of storage required at a machine depends on the cycle times of the pallets (the benefits of mounting a number of parts on one fixture to increase cycle times have already been covered in Chapter 4), the number and type of the work-handling carts, the number of machines they have to serve, and the frequency of that service. Other factors that influence the decision of the amount of machine storage to include are the space available to accommodate the system, the maintenance schedules needed for the work-handling system, and the shift pattern to be worked by the pallet loaders.

There are three main types of machine shuttle. Their configurations are shown diagrammatically in Fig. 5.3. Fig. 5.3a shows a linear shuttle arrangement, which provides two spaces on either side of the worktable position for pallets, a 2+2 arrangement. This arrangement is used by a number of systems, including the Allis Chalmers system, which was described in Chapter 3, and the Vought system, illustrated in Fig. 1.1. Pallets are delivered to one end of the shuttle and collected from the other so that delivery and collection are independent. A 2+2-position linear shuttle can be used for temporarily storing empty pallets if needed

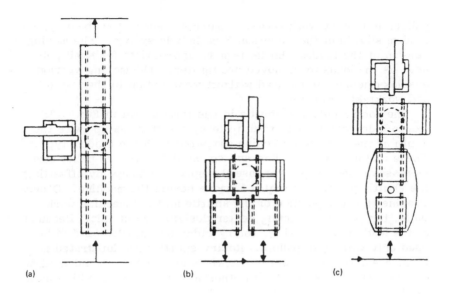

(a) (b) (c)

Fig. 5.3 Types of machine shuttle: (a) linear, (b) tandem, (c) rotary indexing.

because the presence of an empty pallet does not block the machine. The empty pallet can pass through the worktable position when the machined workpiece is completed because there is room for two pallets on the unload side of the shuttle.

Figures 5.3b and c show shuttle arrangements in which pallet exchange only occurs from the front of the machine. They will be termed a tandem shuttle and a rotary indexing shuttle, respectively. The shuttles shown can hold two pallets but have to keep one location empty for when they have to exchange pallets with the machine. Versions of the rotary shuttle exist that have four pallet positions at 90 degrees.

Two versions of the tandem shuttle exist. In the one shown in Fig. 5.3b, there are two fixed pallet plinths (also called pallet stands), and the machine worktable is given sufficient travel so that it can move to be opposite either plinth for pallet exchange. The alternative to this is to have the plinths moving by having them mounted on a traverse slide so that pallet exchange takes place with the machine worktable in one position. The fixed tandem shuttle tends to be used when a system has large

pallets and heavy workpieces. Figure 4.1 shows the fixed-type
tandem shuttle of the Anderson Strathclyde system. The moving
version of the tandem shuttle is more appropriate for small pallets, as the loads to be moved are lighter. The moving version
enables machines to be used without modification to their standard worktable traverse.

The rotary indexing shuttle in the form shown in Fig. 5.3c
has to index carrying two full pallets, and it takes up twice the
depth at the front of a machine compared with the tandem type.
These factors create loading and space difficulties as pallet size
increases. Some suppliers ease the space difficulty by offsetting
the shuttle position so that it is more beside the machine. Others
combine the rear position of the shuttle and the machine worktable, but not all machining center designs permit this. Because
of the loading on indexing, rotary indexing shuttles tend to be
used only with small pallets. Rotary shuttles are illustrated in
Figs. 5.4 and 5.5. It will be shown later how rotary and tandem
shuttles can be combined with other forms of work-handling equipment

5.3.1 System Storage

Shuttles do not have to interface directly to work-handling carts
in all cases. They can interface to localized storage. Figure 5.4
shows a rotary shuttle in the center of some stand-alone pallet
plinths, the shuttle being able to transfer any of the pallets to
the machine. The arrangement shown only requires one of the
stands to be accessible to some other form of work-handling equipment for the arrangement to form part of an FMS.

Buffering and storage has only so far been mentioned as providing a means of decoupling the machines and the main work-
handling system. A frequent further requirement for system
storage is to decouple the load/unloading activity from the remainder of the system. This is required when a system is to operate with the load/unload being unmanned for periods. As with
the original Molins System 24 concept, sufficient system storage
is then required to store enough fixtured workpieces on their
pallets to keep the system supplied with parts during the unmanned period, and to store machined parts. This storage can
be provided by:

1. Individual pallet plinths
2. Racking

Fig. 5.4 Machining module comprising HMC served by rotary in-dexing shuttle and pallet plinths. (Courtesy Kearney & Trecker Corporation.)

3. Carousel storage units
4. Automatic storage and retrieval systems (AS/RS)

Individual pallet plinths come in all shapes and sizes, depending on the system. Examples are illustrated in Figs. 5.4, 4.9, 4.4, and 4.1. Figure 4.1 shows plinths that are also part of a tandem shuttle. The use of racking will be discussed later; an example is illustrated in Fig. 5.10. A carousel storage unit is illustrated on the layout shown in Fig. 1.1, and a small form of the device is shown in Fig. 3.9. Its distinguishing feature is that the stored pallets move around and can, therefore, be transferred on and off the carousel at a single point. Such units are naturally limited in size, as the loads on the drive units increase as the size of the pallets increases.

System storage is often located at the load/unload area, as the storage units (whether plinths or carousels) can double up

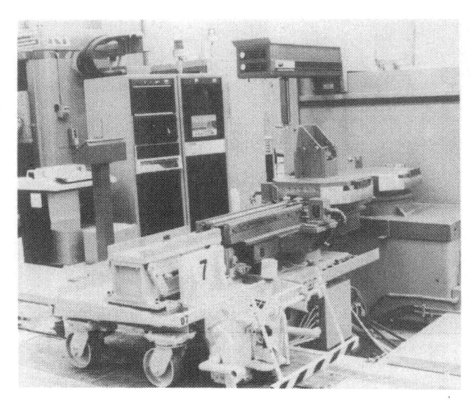

Fig. 5.5 Tow cart with rotary indexing shuttle serving a coordinate measuring machine. (Courtesy Hughes Aircraft Company.)

as pallet supports for load/unloading to be carried out. However, although carousel units are shown at two load/unload positions in Fig. 1.1, they could equally be positioned at machines or elsewhere in the system. It is interesting to note that a five-machine system installed at Leyland Bus, in England, has 42 pallet plinths, these being used to permit unmanned operation for 4 hours in every 12 hours. The pallet plinths in this system are separate stands, each being accessed individually by an AGV work-handling system.

5.4 TOW CARTS

Tow carts were the form of work handling used on the first FMS, the Kearney and Trecker system built for Allis Chalmers. The

Allis Chalmers system was described in Chapter 3; an example of a tow cart loaded with pallet, fixture, and part is illustrated in Fig. 3.7.

Tow carts consist basically of a simple platform on four castored wheels. Pallets are pushed onto or pulled from the side of the carts, when a cart is alongside a machine shuttle. Carts have no suspension that can be variably compressed by the loads they are carrying, so that the cart platform is always at the correct height for pallets to transfer onto or from a machine shuttle. The features of a cart, without a pallet, can be seen in Fig. 5.5. This cart is shown alongside a rotary shuttle serving the coordinate measuring machine in the Hughes Aircraft flexible fabrication system (FFS) [4]. The layout of the Hughes Aircraft system is shown in Fig. 5.6.

Tow carts have a tow pin at the front which picks up on links of continuously driven chains that run under the cart in the floor. A continuous slot provides access to the chain. Carts are controlled from the floor by floor-mounted devices. To stop a cart, a cam is raised from the floor, and this in turn raises the drive pin of the next cart to pass over it and thus stops it. If a cart has to be positioned accurately when stopped so that pallet transfer can take place, then a floor-mounted cylinder pushes the cart against a positive stop which is raised with the pin-disengaging cam. The cams and stops provide one element in the control of a tow cart system. Tow cart systems are one-way systems, and they are controlled in a similar way to railways. The track is divided into a number of control zones, and only one cart is allowed in any one zone at a time. The number of carts in the system must thus be less than the number of control zones, but a reasonably sized system may well have 20 carts in it.

Tow carts provide few restrictions on the layout of a system, unlike conveyors, which generally dictate a straight-line or rectangular-loop layout. The layout variations possible can be seen by contrasting the layout of the Hughes system, illustrated in Fig. 5.6, with the layout of the Allis Chalmers system, illustrated in Fig. 3.6. The Hughes Aircraft system has machines arranged on both sides of two aisles, with the carts only visiting the front of the machines, whereas the Allis Chalmers system has two rows of machines, with the carts passing on both sides. With both systems, variable routings around the one-way system are provided by points. Carts need a certain minimum turning radius, and this can be seen to be provided on the Hughes system without taking excessive room and the minimum radius of curvature needed is quite small.

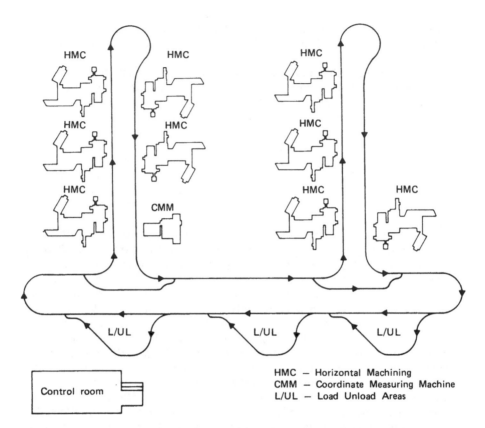

Fig. 5.6 The layout of the Hughes Aircraft flexible fabrication system.

The second system Kearney and Trecker supplied (in the early 1970s) was to Rockwell International at Newark (OH). This system had to fit into a comparatively small area in an existing shop. A tow cart system was implemented satisfactorily, and because of Kearney and Trecker's experience with tow carts, other forms of work handling might not have been considered. However, looking at the forms of work handling available, it is doubtful whether many could meet the requirements of the system within the space available.

It can be seen from Figs. 5.5 and 3.7 that each cart is visibly numbered. The cart will also be mechanically coded with this number, and the number will be automatically read at various positions as the cart moves around a system. The pallets the carts carry are also coded, and this code will generally be read and checked before a pallet is transferred onto a machine. Various means of coding and reading codes are available. Although this topic is mentioned under the tow cart heading, pallet coding applies to whatever work-handling system is used. The transport device will be coded in all systems that have a number of transport units.

The great advantage of tow carts is that they are simple. Their simplicity is derived from their having no on-board power or means of control. This results not only in carts that are fairly cheap, but also carts that are small and need far less floor space than most forms of powered cart. The low cost of carts means they are not a critical resource, and a substantial number can be used. Tow carts can carry substantial loads, but they usually require a steel track to be embedded in the floor to take the wheel loads. Although tow cart systems provide an unobstructed floor, they do require a good deal of floor preparation because the steel track, the power, and the control devices are all mounted in the floor.

Kearney and Trecker have supplied a number of systems using tow carts, and users have found them satisfactory. Maintenance problems have arisen, but considering the duty imposed on the carts and the loads some of them carry, this is not surprising. Most of the longest established systems and also the systems with the largest numbers of machines have tow carts for work handling. These statistics speak for themselves.

5.5 RAIL CARTS

The first FMS with rail carts was the system supplied to Caterpillar Tractor by Sundstrand in about 1974 [5]. Rail carts are generally larger than tow carts because they have on-board drive motors and control systems that have to be accommodated. The typical layout of a rail cart system is shown in Fig. 5.7; Fig. 5.8 shows part of an actual system with a medium-size cart. Figure 4.4 illustrates the load/unload and pallet plinth storage areas adjacent to a rail cart track, such as is illustrated by the layout in Fig. 5.7. Machines can be seen at the far side of the track and at the end of the pallet storage area. Figure 4.13 illustrates the GE, Erie, system, which has a rail cart.

Rail cart systems must have straight-line layouts because the permissible radius of track curvature is so large as to be not

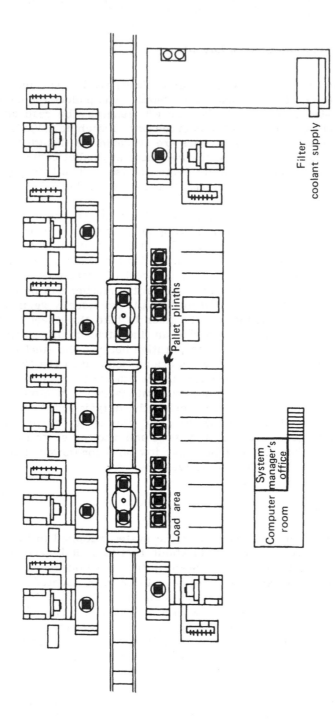

Fig. 5.7 An FMS layout using rail carts with indexable pallet carriers.

Fig. 5.8 View down a rail cart line. (Courtesy Yamazaki Machinery UK Ltd.)

worth having. Although rectangular loop layouts could be designed, in practice rail carts move in both directions on a single straight-line track. They thus differ from most other types of work-handling equipment in that they are two-way systems. This naturally restricts the number of carts that can be in a system; two carts is the usual maximum, with a single cart being fairly common. In a two-cart system, each cart typically serves one-half of the system with an overlap in the middle. The load/unload area is generally positioned centrally.

The operation of a rail cart over its long traverse may be compared to the operation of a single axis of a machine tool, which also has a long traverse. Carts are often rack-and-pinion-driven, with the power to drive the pinion and all the control instructions being passed to the servo drive system on the cart through an attached "caterpillar" cable or by means of a busbar. Sometimes the control instructions are transmitted through an RF signal. Other means of communicating with moving vehicles are under development, and the technology is changing quite rapidly.

In the early applications, rail carts were a very practical solution to the problem of carrying heavy loads about on floors that could have a relatively inexpensive preparation. Although this is still true, the application of rail carts is no longer, in practice, restricted to carrying heavy loads. Most rail cart systems have the rails laid on the floor, so that the rail area is effectively dedicated to the cart. Some systems now have the rails and rack laid in the floor so that the surface of the floor is not obstructed. Some other systems, such as the first Caterpillar Tractor system, have the rails supported on a structure some distance above the floor. As with conveyor systems, this arrangement completely obstructs the area between the machines.

An interesting feature of the design of some rail carts and also of some of the other vehicles used in FMS is illustrated in Fig. 5.7. It will be noticed that the system machines do not have shuttles between them and the rail carts. There are, however, two rotary indexing "shuttles" in the system, but these are mounted on the rail carts. These are only shuttles in the sense that they have the same physical configuration and are able to swap pallets with a machine or a stationary pallet plinth. They are more accurately described as rotary indexing pallet carriers. It is equally possible to have rail carts with two pallet carriers in a tandem configuration, with either pallet position being able to exchange a pallet with a machine or plinth. This removes the requirement for a shuttle at each machine, but it usually requires other storage somewhere in the system, such as the pallet plinths illustrated in Fig. 5.7, and it does not decouple the operation of the work-handling system from the machines. With these arrangements, the carts have to position themselves with a new pallet by a machine prior to the end of a machining cycle if machine utilization is to be kept high. Whether this is a viable means of operating can only be determined by simulation studies. It will be apparent that the longer the machine cycle times, the more feasible this mode of operating becomes. For a system of the size and configuration shown in Fig. 5.7, it has been found that as long as the average machining cycle time is above 40 minutes and the minimum cycle time is not below 15

minutes, then the arrangement can work without adversely affect-
ing machine utilization.

Few features of FMS hardware are static in their development,
and as more manufacturers enter the FMS market, the variety of
hardware design increases. This is particularly true of the design of rail carts. These now come in many shapes and sizes,
with different drive systems and means of transmitting instructions. Some of these techniques have been borrowed from developments that have taken place in the technology of automatic
guided vehicles (AGVs), considered in Section 5.6.

5.5.1 Rail Trolleys

What will be termed rail trolleys are included here after rail carts
because they also have wheels and run on rails. The similarity
stops there, however. A small rail trolley on its track is illustrated in Fig. 4.7. Rail trolleys are distinguished from rail carts
by being externally driven and controlled. Trolleys are thus
comparatively simple in design. Two common forms of drive are
by the use of a chain running down the center of the track or by
the use of a drive tube, as in the Cartrac systems supplied by
SI Handling and illustrated in Fig. 4.7. The drive tube rotates
against a drive wheel on the underside of the trolley, and together they act as a pair of cross-axis rollers, imparting longi-
tudinal motion to the drive wheel on the trolley as the drive shaft
rotates. Such trolleys are controlled by whether or not the drive
tube is rotated and by the angle of the axis of the drive wheel on
the trolley.

Rail trolleys are similar to tow carts in being powered and controlled from their track. Because they have wheels and run on
tracks above the floor, they also have features of the wheeled
pallets, which were discussed in Chapter 4. Smaller rail trolley
systems also resemble conveyor systems in two ways: first by
having a substantial fixed structure mounted on the floor and
second by being laid out in rectangular loops. For example, the
layout of the trolley system of Fig. 4.7 is very similar to the layout of the Sundstrand Aviation, Borg Warner, and SCAMP conveyor systems already discussed. In all four systems, long, thin
rectangular loop layouts are used, and trolleys or pallets circulate one way around the loop. The machining and other workstations are positioned beside the long sides of the rectangle, and
turntable devices at the ends connect the two long sides, thus
enabling the circulation of pallets or trolleys to take place.

The advantages and disadvantages of rail trolleys are comparable to those of the other types of work-handling equipment with which they have been compared.

5.6 AUTOMATIC GUIDED VEHICLES

The most recent vehicle to be used in FMS is the automatic guided vehicles (AGV) — also termed wire-guided vehicle (WGV). It is a reasonable generalization to say that, in terms of their application to FMS, the 1960s saw the first use of the conveyors for work handling (in the early integrated systems), the 1970s saw the start of the use of tow carts and rail carts, and the AGV is the vehicle of the 1980s [6]. It is, however, worth noting that a form of WGV was used in the Herbert Ingersoll plant in 1968 to link different parts of the plant (see Fig. 3.3). A typical modern AGV is illustrated in Fig. 5.9.

AGVs are self-powered, electrically driven vehicles, and they carry a substantial battery pack to power their drive motors. Most AGVs have two independently driven drive wheels mounted on either side in the middle of the vehicle. Steering is achieved by driving the drive wheels at different speeds. Because the drive wheels are central, they also carry most of the load. The front and back of the vehicle are supported by one or two castor wheels. AGVs can also have a tricycle design, with a single steering and driving wheel at the front and two passive wheels at the rear. Most AGVs pick up their guidance and control instructions from a wire buried in a small trench in the floor. A low-frequency current is fed to the wire, and this is inductively picked up by sensors on the vehicles. An on-board computer interprets the signals and controls the vehicle.

An AGV layout is shown in Figs. 5.10 and 1.1, and it can be seen that a number of paths can be followed. The paths to be followed and the individual vehicles can be distinguished by the controlling computer by using more than one wire and more than one frequency. In addition, sensors and transmitters can be positioned in the floor to communicate with a vehicle as it passes over or waits above a communication point. This communication can include programming a vehicle to carry out a series of movements and to proceed to a further programming point once the movements have been completed.

Other means of communication with and controlling AGVs have been developed and are under development [7]. These include systems that have the vehicles follow white-line, chemical, or

Fig. 5.9 An AGV. (Courtesy Cincinnati Milacron.)

magnetic guidepaths, which can be simply deposited on the floor rather than buried in it, and others that use a "virtual guide-path." The latter methods provide means for vehicles to deduce their own positions and their own paths. The methods can in-clude dead-reckoning systems, in which sufficient information is stored in the on-board microcomputer for an AGV to plan its path between known positions; systems having beam-emitting beacons, which enable an AGV to navigate itself; inertial-navigation sys-tems, in which an AGV measures its own movements and uses this information to work out its position; direct imaging, where cameras or ultrasonic scanners enable an AGV to "see" where it is and hence plan its path to a destination. The established

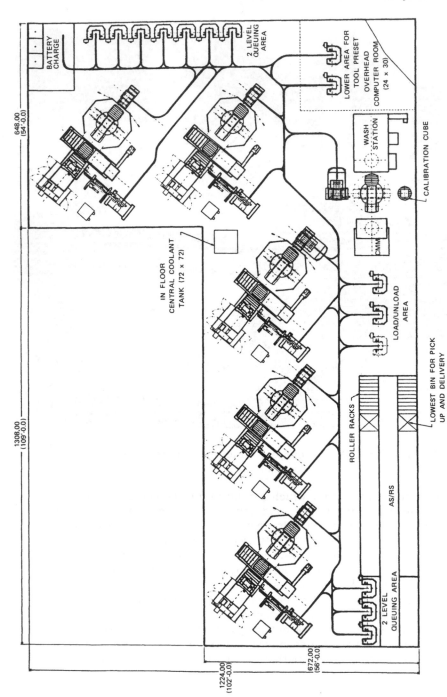

Fig. 5.10 An FMS layout using AGVs. (Courtesy White Sundstrand, WCI.)

technology is, however, the wire-guided vehicle, and this is found
in most AGVs used in FMS.

AGVs cannot guarantee to position themselves against machine
shuttles as accurately as tow carts or rail carts. Inaccuracies can
arise in both horizontal and vertical alignments. Vertical misalign-
ment arises because AGVs have rubber-drive wheels and a degree
of suspension to ensure their wheels have a positive grip on the
floor. In the lateral directions, an AGV's guidance system cannot
be expected to compete with the accuracy provided by rail guidance. AGVs thus have to be located positively at pallet transfer
points. Many AGV systems achieve accurate location by lowering
either the whole vehicle or the pallet-carrying platform onto metal
cones either fixed on the floor or mounted on an adjacent machine
structure. Such an arrangement can be seen in Fig. 5.9. The lateral arms projecting from the top of the AGV have the female cones
on their undersides; the male cones are visible by the machine to
the left of the figure. A similar arrangement can be seen in Fig.
4.5, with pairs of cones being clearly visible on the sides of the
pallet plinths. Once the AGV is located, the pallet can be transferred either off the front or to the sides of the AGV.

An alternate means of location is for the AGV to drive into a
docking stand which has slightly converging sides and a positive
end stop. The pallet carrier may still move vertically with this
configuration, but the vertical movement is used to deposit the pallet onto the machine shuttle or a pallet plinth. The AGV then re-
verses out. This pattern of delivery and pickup is used in the systems illustrated in Figs. 5.10 and 1.1. In Fig. 5.10, a number of
pallet plinths of the docking-stand type can be seen in the queueing areas and in the areas for tool setting and load/unloading.

AGVs provide very little restriction on system layout as long
as adequate space can be provided for the radii needed for turning corners. Figure 1.1 shows an AGV system that was laid out
in a shop which had been cleared to receive it. The machines are
arranged in two parallel rows, and most of the AGV movements are
in two perpendicular directions. The system shown in Fig. 5.10
provides a complete contrast to this in that the system has been
fitted into an L-shaped piece of restricted width. By placing the
machines at an angle and by taking advantage of the reversing
and turning capabilities of AGVs, the system has been fitted into
the space available. The AGV paths of Fig. 5.10 all come off a
single spine, and this is unusual. It is more usual to have a series of unidirection loops connected together, rather like tow cart
layouts. Space restrictions makes this impractical in the system
illustrated.

In addition to facilitating irregular layouts, AGVs have the advantage of leaving the floor areas around the machines relatively

uncluttered. The floor used with an AGV does, however, need
to be carefully prepared, and this can be quite expensive. With
wire-guided AGVs, there are limits on the permissible metallic
content of the floors, as embedded metal can interfere with the
signals emitted by the wires. Even with well-prepared floors,
the rubber wheels of AGVs can skid if the floor gets coolant or
other dirt on it and the tires can also pick up swarf. AGVs tend
to be large and heavy because they carry a heavy battery pack
even before being loaded with pallet-fixtures and parts. AGVs
are thus larger and heavier vehicles than the equivalent tow carts
to carry the same load. Their mass and momentum when moving
limits the safe speed that they can travel unless they are in a
fully protected area.

As with all other types of work handling discussed in this
chapter, AGVs were applied in FMS, not developed for FMS. The
number of AGV applications in FMS is comparatively small compared with the total and rapidly increasing number of applications of AGVs in areas such as assembly, warehousing, and general handling. The development engineering taking place to support these other applications will benefit FMS applications and help
to keep AGVs competitive with other types of work handling.

5.6.1 Forklift AGVs

All the work-handling systems so far discussed have moved pallets about throughout the system in one horizontal plane, and
this has been the plane at the height of the machine worktables.
The storage carousels, load/unload stations, machines, and shuttles have all been at this height. This aspect of system design
has actually simplified the mechanical design, as pallets simply
have to be pushed or pulled on roller slideways when being transferred between work handling, shuttles, plinths, machines, and
so forth.

Figure 5.11 illustrates a forklift AGV which, by virtue of its
mast, adds another dimension to a system. The benefit of the
forklift AGV is illustrated in the figure. It can access vertical
storage racking, and this makes it possible to use standard multitiered racking for storing loaded pallets. This racking is much
cheaper than either using individual plinths for each pallet stored
or using storage carousels. It also reduces floor space requirements, depending on the amount of storage needed and the height
of the racking available. Extra space can be needed, however,
because the length of the forks increases the length of the AGV,
and it is necessary to have extra space in front of the racking

Fig. 5.11 A forklift AGV accessing storage racks. (Courtesy AB Bygg-och Transportekonomi, BT.)

to align the AGV prior to picking up or delivering a load. Telescopic forks reduce this requirement.

Forklift trucks traditionally carry out many handling tasks within factories. Their forks are a standard and flexible interface to wooden pallets or bins. A forklift AGV can therefore easily be used to carry out a variety of tasks, some of which may be as part of an FMS while others may be general handling tasks. The forklift AGV can thus be a means of linking an FMS into other elements of the operation of a factory. This is also true of the next form of work handling considered.

5.7 STACKER CRANES

Stacker crane is a generic term that refers to a variety of devices used in the aisles of the racks of warehouses to access the

individual storage locations. They have a vertical mast and a
carriage that travels up and down the mast to access any of the
storage locations on both sides of an aisle. There is a variety
of designs of stacker crane. In the design from which they probably get their name, the vertical mast hangs from the carriage of
an overhead crane. The gantry and carriage of the crane permit
two-directional lateral movement of the mast, so that the mast is
not dedicated to a single aisle of racking. Equally, the crane is
not confined to the aisles of the warehouse, but can bring parts
from the warehouse directly into processing areas without intermediate work-handling devices being required.

A second design of stacker crane dispenses with the overhead
gantry and instead has parallel monorails that support the top
and bottom of the mast, the bottom monorail providing the main
support. This is the more usual form of stacker crane used with
FMS. An example is shown in Fig. 5.12. In older forms of stacker crane, the vertically moving carriage carried an operative who
did the picking of parts from the storage locations and controlled
the movement of the crane. Modern forms of stacker crane are
computer-controlled, and those in automatic storage and retrieval
systems (AS/RS) have automatic means of accessing the storage
locations and bringing bins out of the warehouse for picking to
take place.

If designed with suitable lengths of monorail to do so, there
is nothing to stop a stacker crane coming out from an aisle of an
AS/RS and delivering items directly to the machines of an adjacent FMS. Other handling tasks, such as transferring parts between machines or to load/unload stations, can also be performed.
They can thus both link an AS/RS to an FMS and provide work
handling within an FMS. Stacker cranes have swiveling forks for
picking up pallets and for accessing rack locations on both sides
of a storage aisle. There are thus similarities between the modes
of operation of stacker cranes and the forklift AGVs, just discussed. The analogy cannot be extended too far, however, because monorail stacker cranes are a form of rail cart and, like
other rail carts, are restricted to moving in a straight line. This
is both a restriction and a benefit. Stacker-crane-based systems
can be very compact, and the stacker crane itself can move from
place to place far more quickly than AGVs. The stacker crane in
Fig. 5.12 is part of a FMS for processing sheet metal parts. The
AS/RS shown holds sheet metal blanks. Sheet metal FMS are discussed in Chapter 6 together with a novel application of an AS/
RS and stacker crane used in a flexible assembly system.

Fig. 5.12 A stacker crane within an AS/RS for sheet metal blanks. (Courtesy Trumpf Machine Tools Ltd.)

5.8 ROBOTS

The use of robots in handling nonpalletized workpieces that can be loaded directly into fixtures on machines was discussed in Chapter 4 when fixtures and pallets were considered. It was then stated that robots generally have insufficient lifting capacity to handle palletized workpieces. Although this is generally true, as with many other statements about FMS, it is usually

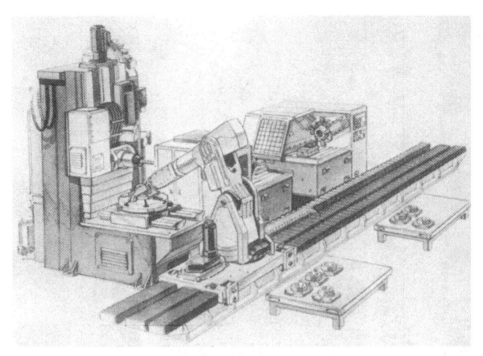

Fig. 5.13 A rail-cart-mounted Smart robot. (Courtesy Comau/ RSLA Marketing International Ltd.)

possible to find exceptions. In this instance, there are at least two exceptions which will be presented. Both illustrate untypical robot applications and the ingenuity of designers in developing different approaches to work handling.

A stationary robot can generally only physically reach one or two machines to load them, and so only low-cost and hence low-capacity robots can be justified for such applications. To provide a link to other machines, an additional means of work handling is required. If the range of a robot can be extended so that it can undertake the work handling required for a group of machines, then it becomes economical to have a heavier-duty and more versatile robot. Examples of this are shown in Figs. 5.13 and 5.14. Figure 5.13 shows a large Smart robot mounted on a single-axis slideway. This is equivalent to a robot mounted on a rail cart, or it may be described as a seven-axis robot. The size of the robot means it is capable of a variety of handling tasks, and the slideway means it can link a number of machines

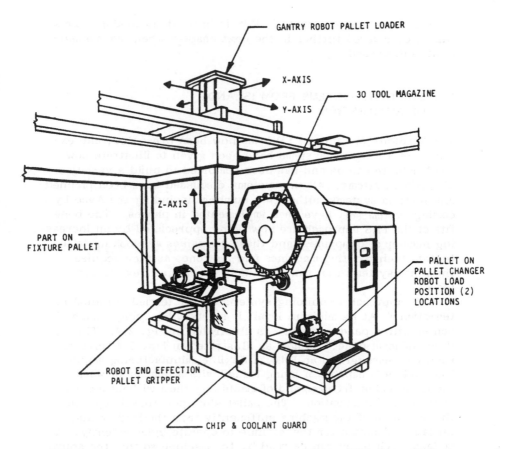

GANTRY ROBOT PALLET LOADER

X-AXIS

30 TOOL MAGAZINE

Y-AXIS

Z-AXIS

PART ON
FIXTURE PALLET

PALLET ON
PALLET CHANGER
ROBOT LOAD
POSITION (2)
LOCATIONS

ROBOT END EFFECTION
PALLET GRIPPER

CHIP & COOLANT GUARD

Fig. 5.14 A gantry robot serving a machine. (Courtesy White Sundstrand, WCI.)

and thus provide the integration needed to form an FMC or FMS. Figure 5.14 shows another means of extending the range of a robot, i.e., mounting it on an overhead gantry. The two axes of the gantry enable a number of machines to be spanned. The floor preparation required is minimized, and the floor can be kept uncluttered. As with forklift AGVs and stacker cranes, the gantry-mounted robot can operate at a number of levels, and as with suspended stacker cranes, it has complete freedom of movement within its operating area. The form of robot illustrated is designed to change pallets on machines.

A growing use for robots in FMS is in handling tooling. This will be considered further in the next chapter when tool management is discussed.

5.9 BUILDING SYSTEMS FROM ISLANDS OF AUTOMATION

The hardware of FMS and FMC has now been presented, and examples of a range of systems have been given to illustrate how the hardware can be and has been combined to build a variety of flexible systems. All the systems discussed have been planned and built as systems, although a number, including the Avco Lycoming system [8], have been implemented in phases. The benefits of the FMS approach are now being appreciated by an increasing number of companies, and these companies are looking to integrate their existing machines and new ones as they acquire them into systems that can eventually operate as flexible systems.

This approach to building systems can be termed "gradual integration." At its minimum level, it starts with a single machining center with a pallet changer, as illustrated in Fig. 2.4. The addition of local pallet storage, as illustrated in Fig. 5.4, creates what has been termed an FMM, a flexible manufacturing module. The module is capable of being loaded with a number of workpieces on pallet-fixtures and of machining them without significant human intervention. The pallet storage carousel separates the operation of the machine sufficiently from the loading and unloading function for the machine to operate independently. A code on each pallet can be read by the machine so that the appropriate machining programs are called up from its CNC memories. The necessary tools have to be in the carousel prior to machining, and this limitation of tooling capacity restricts the machining that can be carried out without the intervention of an operator.

The addition of a second similar machine can start the progress of developing from the module toward a flexible system. The machines must be linked by one of the many means of work handling described, and pallet loading and unloading must be centralized so that workpieces can be scheduled to either machine. Figure 5.15 shows such a two-machine cell. In this example, the work-handling link is provided by a rail cart with the rails secured to the top of the existing floor. The addition of the work-handling link must bring with it a supervisory

Fig. 5.15 A two-machine rail-cart-served FMS. (Courtesy Kearney and Trecker Marwin Ltd.)

control system to control the work-handling and the scheduling of the workpieces to the machines and to storage devices. Features of this software will be considered in Chapter 6. A rail cart is an obvious choice for linking a limited number of machines, as extensions to the rails (to link to further machines) are easily made without a great deal of attention having to be paid to the floor. In addition, a two-way cart does not require the creation of a loop path (and hence require more floor space), as would the use of a tow cart for example.

The scenario just described suggests a rather piecemeal approach, which is not the ideal way to plan such an integration. Although "gradual integration" can be implemented step by step, it is not the recommended way for the integration to be planned.

The step from a single module to the linking of a second machine
has all the features involved in linking further machines, and it
is better to consider how the third, fourth, and fifth machines
are to be integrated at the same time as planning for the second.
It is also advisable to plan beyond the island of automation that
will be created by linking a few machines to see how a totally integrated manufacturing system can be designed. This total approach to manufacturing system development is the one that has
to be adopted to obtain maximum benefits. It is, however, the
most demanding for companies. The creation of a single large-
scale, integrated system at one time can overstretch a company,
in terms of both its financial resources and its engineering resources. Also, the larger the system created, the longer it takes
to implement, the longer to tune to maximum operating efficiency,
and the longer to obtain a financial return. Thus, system development often has to proceed through the creation of islands of automation, as just discussed. These can be fully commissioned and
start earning their keep relatively quickly and provide both the
building blocks for larger-scale integration and the cash to fund
it.

System integration does not always involve complete automation. The seven-machine Gardner FMS, discussed earlier in this
chapter, was shown to be part of a larger 14-machine system that
operated as a unit. The control procedures available from such a
system can often be used to govern what happens both upstream
and downstream of it without complete integration or automation
of all the machines being required. All that is required is that
the utilization levels planned for the associated machines and
processes are less than those planned for the integrated system,
so that the associated machines never starve or block the FMS.
The nucleus of such a system is the high capital cost FMS, and
thus it is only really in this part that it is important to keep
utilization levels high.

This approach to integration has been applied very effectively on a small scale by Normalair-Garrett Ltd. (NGL) at their
Crewkerne, England, plant. NGL is a major supplier of control
systems and component assemblies for aerospace, defense, marine,
and allied industries. Its integrated system was designed as a
manufacturing facility for subassemblies for aerospace applications. NGL analyzed the subassemblies it was required to produce and found that 85% of the manufacturing costs were incurred
in the manufacture of a small number of complex prismatic parts.
NGL therefore set out to control the costs, stocks, and work-in-

progress involved in the manufacture and assembly of these parts by an integrated approach that incorporates both flexible and just-in-time manufacturing.

The system is based around a two-machine FMC which has two KTM machining centers linked by a rail cart. To complete the system there are a CNC lathe, a spark eroder, a milling and a grinding machine, heat treatment and cadmium-plating facilities, and an assembly and test area. The FMC is used to machine the complex prismatic parts, and sets of assemblies of parts are machined at a time, as required by assembly. This can be done quite literally because the parts are mostly small and multipart fixtures are used. Although this makes for rather complex fixtures, the principle of machining by subassembly sets was fundamental to the approach adopted.

Another principle of operation is that the high-cost FMC is kept busy and high utilization is maintained, whereas lower utilizations are quite acceptable on the lower-cost stand-alone machines. Labor flexibility permits effort to be applied where it is most needed. The FMC control is linked into an integrated management control system for the manufacturing facility, and this in turn is linked through to the company's mainframes. Thus, although the JIT approach means that assembly requirements determine what the FMC produces, what the FMC produces links through to all the other machines and processes up to and including final assembly.

NGL designed and implemented much of its total system itself, including some of the software. More companies are now adopting this approach to ensure they have the in-house expertise to effectively operate and further develop their systems. This approach is now facilitated by a number of suppliers who are packaging their hardware and their software [9] so that smaller companies can implement flexible systems through stages. Some independent software houses and computer manufacturers are now also interested in supplying software to integrate machines and work handling into systems, and this will further enable gradual integration by companies. The software and control of systems is the first topic in Chapter 6.

REFERENCES

1. M. Atkey, "Manchester FMS unites 14 machines," *Machinery and Production Engineering*, July 21, 30–33 (1982).

2.   M. Lincoln, "How Borg Warner tackles FMS," *Production Engineer*, March 12–13 (1984).

3.   Anon., "Today's FMS, Comau," *Iron Age*, Aug. 16, 39 (1985).

4.   P. M. Burgam, "Flexible fabrication moves in at Hughes Aircraft," *Manufacturing Engineering*, September, 56-57 (1983).

5.   Anon., "DNC for flexibility," *Production*, September, 70–77 (1974).

6.   R. L. Hatschek, "Guided carts link machines into systems," *American Machinist*, August, 97–100 (1980).

7.   S-E. Anderson (ed.), *Proceedings of 3rd Int. Conf. on Automated Guided Vehicle Systems*, Stockholm, IFS (Publications), Bedford, UK, and Elsevier Science Publishing Company Inc., New York, pp. 101–218 (1985).

8.   Anon., "Our 'FMS' will do the work of 67 conventional machine tools," *Production*, April, 66–69 (1978).

9.   T. P. Shifo, "A standard product approach to FMS," *Commline*, *XII*(6), 16–20, Nov/Dec (1983).

6

System Management and the
Developing Scene

6.1 INTRODUCTION

Designing an FMS is a multifaceted task, part of which involves
a system vendor and a customer collaboratively selecting the appropriate hardware and planning how the system will operate.
Assumptions concerning the operation of the system will reflect
how it is seen that the system will be managed and controlled.
The scenarios of possible operation will be investigated by modeling and simulation techniques, and these are presented in subsequent chapters. What is modeled and simulated is dependent
partly on the hardware selected, partly on the control procedures that are built into the system software, and partly on how
the system is to be manned and managed. Details of system
hardware have been presented in Chapters 4 and 5. Aspects
of how systems may be managed and controlled are covered in
this chapter.

It was stressed in earlier chapters that the book would concentrate on metal-cutting FMS because all the early systems and
most of the new systems are metal-cutting FMS. It would be

inappropriate to conclude this section of the book, however, without again putting flexible manufacturing into a larger manufacturing context and showing how the ideas behind flexible manufacturing are finding wider application and the technology associated with FMS is developing. These topics are covered in the last section of this chapter.

6.2 THE CONTROL OF FMS

The prime element in the integration of any system of machines and work-handling equipment is a computer. Over the 15–20 years that integrated systems have been developing, computers have been undergoing a transformation which has involved an exponential growth in the computing power per chip coupled with progressive miniaturization. This, in turn, has led to larger-capacity computers operating at greater speeds. The growth in capacity and speed is predicted as continuing for the foreseeable future.

The use of computers has always required appropriate combinations of computer hardware and software. While the speed of hardware development has been rapid, the development of software has invariably lagged behind. The reason for this is that most software has to be handwritten and then proved and debugged by its author(s). As the capacity of computers has increased, so their ability to accept and run larger and more complex programs has grown. However, there have been few developments to aid the speed of writing the programs to exploit the increased hardware capacity available. As far as FMS are concerned, the programs required to control a system have taken a considerable number of man-years to produce, and refinements to the programs that run today's FMS are still being developed. Before looking at the refinements, the form of the basic control software as used on early systems will be reviewed.

Early systems generally had two computers to control them, a DNC computer and an FMS computer. The division of control between the computers was as indicated in Fig. 6.1. The FMS computer looked after the control of the work-handling system, the scheduling of pallets to the various machines, and its software kept track of and controlled the movement of the workpieces in the system. The computer maintained a file of pallet-fixtures so that it knew which pallet-fixtures were where in the system with which workpieces and which pallet-fixtures were available to

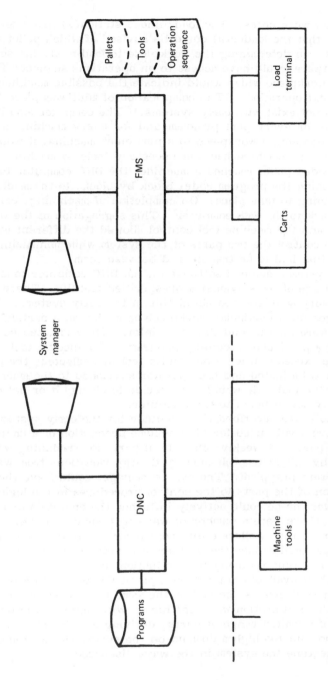

Fig. 6.1 Block diagram of early FMS control software.

accept new workpieces. The load/unload station(s) had a terminal(s) so that the loader(s) could be advised of which pallet to load next. In determining the workpiece to load next, the scheduling software would have access to an operations sequence file for each workpiece which would indicate the possible machines for the next operation. (The complication of multiworkpiece fixtures did not exist with early systems.) The computer also had a tool file for every part program and for every machine and before scheduling a workpiece to a particular machine, it would check that the machine had the necessary tools to machine it. Once a loaded pallet reached a machine, the DNC computer took over, passing the program code, block by block, to the machine for machining to take place. On completion of machining, control passed back to the FMS computer. This segregation of the work-handling and the machine tool control allowed the different computers to control the two parts of the system while minimizing the data that had to be transferred between them.

Most systems included software on the DNC computer to monitor the usage of tools against a prespecified tool life. There was nothing very sophisticated about this in the early systems. The tool usages for a particular workpiece operation on a particular machine were known, and thus this information was accessed on selecting a particular machining program. If a specific tool usage was to be exceeded when a particular tool was selected, the program would be halted and the operator summoned to the machine to check the tool. He would then decide to change it or not and report any action taken to the computer.

The software described was adequate for the early systems, which were mostly used for high-volume production of a limited range of parts. A weekly schedule of parts for machining was planned by the system manager, and large variations from week to week were untypical. The system manager could leave the production of the parts to the control procedures in the logic of the software or he could actively influence the operation of the system. There were a number of ways this could be done. He could, for example, allocate different priorities to different workpieces, he could reduce the number of pallets available for a particular part, thus reducing its production, or he could restrict the machines available to produce a particular part. In making any of these decisions, he had at the same time to try and ensure machine utilization was kept high. With high-volume production of a limited range of parts, deriving a satisfactory schedule did not put too high a demand on the system manager and he could fine-tune the system in the ways discussed.

6.2.1 Advanced Control Systems

Modern FMS are generally purchased to machine a greater range
of parts than was required of the early FMS. The first effect of
more parts being put through a system is that the "volume" of
control required for organizing the pallets, workpieces, and tooling increases. Second, the increased part variety means that the
system manager no longer finds it easy either to schedule the system manually or to influence its operation effectively. Operating
experience is difficult to gain in an operating environment with a
large variety of parts because operating conditions are continually changing. Thus, the system manager has to look to the FMS
software to control the system more effectively, but, when required, the software must still be responsive to his needs. A
third aspect of increased part variety stems from the reason for
the variety — i.e., systems are being used in lower-batch-quan-
tity environments. In some cases their application even extends
to what may almost be considered jobbing shops with production
of individual parts — a batch size of one. The flexibility needed
from an FMS is thus increased, and tool management becomes a
demanding problem.

In these applications, it is unwise for an FMS to be treated as
a production unit on its own, but it should be integrated into a
larger system, as discussed at the end of Chapter 5. To achieve
this, it is necessary for the control of FMS to be integrated within a larger system of control computers. Figure 6.2 shows a diagrammatic representation of an FMS control system which is drawn
to illustrate both the hardware and software configuration of the
FMS and its integration into the control hierarchy of a company.
The diagram illustrates only the control hierarchy with respect
to one FMS (or production cell), whereas the factory mainframe
computer would also service purchasing, production control, the
stores, accounting, and so forth, in the established data-processing role of a factory mainframe. In addition to supporting
these company functions, the mainframe could also be linked to
other company FMS or computer-controlled processes through
their own control computers.

Under a distributed computer system as illustrated in the figure, there are many possible hardware configurations. For example, each of the boxes at the "functional" level could be a separate computer each controlling its own hardware. Alternatively,
the boxes could just be representative of software mounted with
the scheduler on the cell computer, but controlling the hardware
indicated at the function level. How the hardware is configured

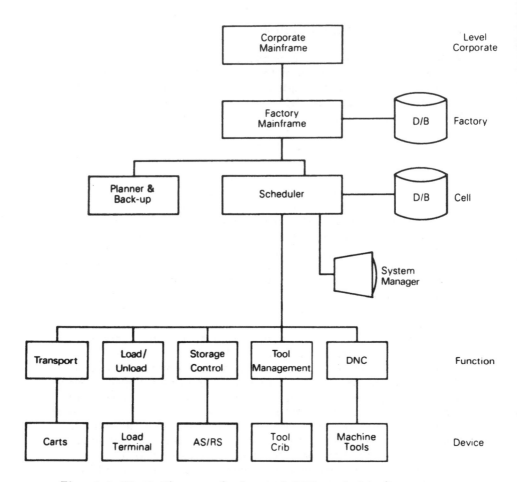

Fig. 6.2 Block diagram of advanced FMS control software.

will depend on the system vendor and who supplies the devices at the lowest level. If a system vendor is primarily a manufacturer of machine tools, then the work transport system and the AS/RS may be supplied by other manufacturers, and these devices will be supplied with their own computer controls. These computers will need to be linked into the system by an interface to the scheduler computer, and this involves extra design work and expense.

A modular design can have benefits if problems arise in another part of the system because the individual device computers enable the devices to keep functioning while problems elsewhere are sorted out. This ability is necessary to meet a design requirement of all systems that a system should be "progressively degradable." This requires that any failure can be isolated to minimize its effect on the remainder of the system and allow the remainder to keep working. How far it is possible to achieve this depends on many design factors, the prime one being the modularity of the control. One part of the system where a progressively degradable approach is difficult to achieve is the cell computer. For this reason, it is not uncommon to have a second system computer to provide backup should the first go down. The backup computer can do more than just provide backup. There is now an increasing amount of decision support software, often based on simulation techniques, which can be run on the second computer to assist the system manager in his task.

With the hierarchy illustrated, only the lowest level of control involves continuous concurrent real-time operation. As the level moves up the hierarchy, the frequency of control reduces as the planning horizon increases. The hierarchy is not only designed to pass control down to the lowest level, it is also arranged to report from the lowest levels to higher levels. How high up the hierarchy any reporting occurs will depend on the information being reported. Most reports on the system operation can stop at the system manager level. The factory computer is really only interested in the production that has been achieved, probably on a daily basis.

The U.S. National Bureau of Standards (NBS) has carried out an extensive research project on the architecture of control systems for FMS and similar types of integrated system [1]. It has designed an architecture that distributes the tasks across a series of levels, as illustrated in Fig. 6.3. The NBS is developing software, interfaces, communication protocols, and data structures to enable an automated manufacturing system to be data-driven using this architecture. Slight differences can be seen between the implementations illustrated in Fig. 6.2 and those in Fig. 6.3. This applies particularly to the NBS architecture having a "shop" level between the factory and the cell.

General Motors has also taken an initiative in this area by sponsoring the development of an international computer communications specification termed manufacturing automation procotol (MAP) [2,3]. MAP is designed to facilitate the form of integration shown in Figs.

Level	Functions performed
Facility (i.e., factory)	Information management Manufacturing engineering Production management
Shop	Task scheduling Resource allocation
Cell	Task analysis Batch management Scheduling Dispatching Monitoring
Workstation	Set up Issue equipment commands Takedown
Equipment (i.e., device)	Processing Measurement Handling Transport Storage

Fig. 6.3 An architecture for FMS. (After Ref. 1.)

6.2 and 6.3 in which the factory mainframe computer may be made by one manufacturer, the intermediate system or cell minicomputers may be made by another, and the device-driving computers may each be made by further different manufacturers. Getting computers made by different manufacturers to communicate with each other has always been very difficult (if not sometimes impossible) because the computers' internal architectures and structures have been completely different. Individual interfaces have been designed as one-off solutions to individual interfacing problems, but these have been expensive. GM recognized that larger-scale integration required the use of standardized networks, interfaces, and protocols. By gaining support from other major computer users, sufficient pressure was brought to bear on computer vendors for them to accept that standards should be agreed and then provided on their products. International collaboration is now taking place to agree on standards for MAP, and recently there have been a number of demonstrations of different vendors' products communicating with each other using standard protocols. It is likely to be the 1990s, however, before all the MAP concepts and specifications are successfully implemented.

Figure 6.2 symbolically shows data storage at the factory and cell levels. Differences in approach do exist here. Companies that plan an integrated approach from the outset have found their database management is easier if all data are held on the factory mainframe and accessed when required. This even applies to NC programs for the FMS machines. Companies that have installed FMS and then linked it up to other computer systems use a more distributed database approach. Their FMS installation has a database to support the operation of the FMS, and this includes workpiece data and NC programs. In these implementations, the link from the factory mainframe is primarily used to provide scheduling information and to report back on completed production. One element of system management and control shown in Fig. 6.2 is tool management. The degree of tool management required depends partly on how a system is manned and how large a variety of parts is to be produced. Aspects of tool management are dealt with after system manning has been discussed.

6.3 THE MANNING OF FMS

In the early and mid-1970s, the term "flexible manufacturing system" was (mistakenly) almost synonymous with "unmanned manufacture." This was partly because the Molins System 24 had unmanned shifts as a fundamental feature of its operation, partly because the Japanese were spending a considerable sum of money on developing unmanned manufacturing with their MUM Project — methodology for unmanned manufacture [4] — and partly because the many advantages of FMS, other than the reduced labor requirements, were not fully understood. Most financial justification procedures only evaluated savings in terms of reduced labor costs, and unmanned operation, almost by definition, was going to offer the maximum saving. It thus seemed natural for all FMS to follow the Molins lead and to have unmanned manufacture as a feature of their operation.

While the Japanese were pursuing their MUM project, various companies in the United States were implementing the early FMS. These were commercial systems built to produce parts — they were not experimental systems and they were not unmanned.

The American users were trying to minimize the labor involved rather than eliminate it. The reasons for this minimal manning approach will be explained in this section as the tasks associated with operating an FMS are discussed. By the end of the 1970s, doubts about the feasibility of unmanned manufacturing were raised when the Japanese MUM project was superseded by a project on manufacturing with lasers without the success of the MUM project being formally demonstrated. However, the Japanese have since achieved successful unmanned operation and have built two factories, one a Fujitsu factory and the other a Yamazaki factory, which operate unmanned in a substantial number of sections.

The reasons why almost all existing systems are currently manned relate to the tasks that have to be carried out during the operation of an FMS. These are described in the following paragraphs.

1. Workpiece fixturing, pallet loading and unloading: The input and output point of a system is the load/unload area. Here load/unloaders unload machined parts, mount new parts in fixtures, and refixture parts that have been partly machined. Few existing systems include a washing unit, so that cleaning of the workpieces, fixtures, and pallets is generally also needed and carried out. Some users also have their loaders carry out workpiece deburring and inspection at the load station. A stock of unmachined and partly machined workpieces is held adjacent to the load/unload area, and this is usually organized by the loaders. Information on which parts to load next will be supplied by an adjacent computer terminal, and when loading is completed, data on what has been loaded must be input to the computer.

2. Tooling maintenance, preparation, and organization: It is usual practice to have all tool preparation, presetting, and maintenance carried out in a tool crib dedicated to an FMS; the work being carried out by a crib operator. The crib operator, having prepared a particular tool, will generally also be responsible for seeing that it gets to the right machine. Without automated tool handling (discussed in the next section), this may involve taking the tool out and placing it alongside a machine ready for use. The crib operator is advised which tools to prepare by the system computer. The information for this is supplied both by the scheduling computer and by the machines' CNC controllers, which monitor tool usage against a preprogrammed tool life specified for each tool. When 75% of the nominal tool life has been reached, the crib operator is advised to prepare another tool.

3. System operation: The tasks to be carried out under this heading are closest to those traditionally carried out by machine operators. The tasks are broader, however, in that the operators attend to the machines and to any aspect of system operation that requires attention. It is thus reasonable to call the operators "system operators."

A major task for such operators in most systems is concerned with tooling. Although the CNC controllers monitor tool usage, technology is only just progressing to a stage where tool wear monitoring (rather than tool usage monitoring) is becoming more practicable. Despite a great deal of research aimed at predicting tool life, many users find that tool life can vary significantly. The prime cause of this is material variability, particularly in castings and forgings. Thus, although a computer can be used to monitor tool usage and warn when, say, 75% of a given tool's predicted life has been reached, it takes an operator to inspect the tool and say whether it has further life or should be changed. The operator can be summoned again at any other tool usage value to make a further tool inspection.

System operators have to be able to operate the machines in manual or manual data input (MDI) mode for setting up new workpieces, correcting faults, or achieving size on a critical bore. In Chapter 4, it was pointed out that machining centers cannot always achieve close machining tolerances by themselves, but they often can with operator assistance. Some users use operators to check critical bores with hand-held air plug gauges and program a stop at the appropriate point in a program to allow for checking. As the bores are generally more often undersize from a tool wearing, the operator makes an appropriate correction to the tool or the offsets and reruns the bore tape blocks before allowing the program to proceed.

All systems have centralized swarf collection and coolant systems. Coolant pressures are kept higher than normal to help move the swarf from the machines into a central flume. Even so, a certain amount of machine cleaning is necessary, and this is another task for system operators, particularly with large machines which cannot be easily guarded.

Systems of the complexity of FMS use a lot of contact and noncontact switches. The movement of any pallet or cart typically necessitates the making and breaking of these switches. The reading of pallet codes prior to machines and at other places requires further use of contact or noncontact devices. These devices operate with perfectly acceptable reliability figures, but they sometimes fail or drift. The devices also rely on pallets or pallet mechanisms maintaining alignment while passing over the devices. Such mechanisms can wear, nuts can become loose, an element can be marginally designed and fail. These are all examples of small system faults which an experienced operator can readily attend to and correct if he is familiar with the system. System diagnostic devices will assist in identifying a particular fault, but it still requires a pair of hands and someone who knows the system to correct the fault.

6.3.1 Manning Levels and Philosophy

In planning how a system is to be manned, it should be remembered that system commissioning can take anywhere from 9 months to 2 years, depending on the size and complexity of the system. The commissioning period should be exploited by the user so that company personnel become fully familiar with the system. All systems thus start off as manned, and manning is gradually reduced as the system comes on stream and as performance and reliability improve.

Most current users have a system manager who works one shift and an industrial engineer who also works one shift. Every five or six machines in a system have one system operator and one or two load/unloaders to handle system input and output. The number of loaders depends on the machining cycle times. Some users use unskilled labor for the loading and unloading, but more adopt the skilled-team approach of rotating the loaders and operators on a weekly or two-weekly basis. The purpose of this is to give the direct-system workers an identity with the whole system and to provide a slightly larger pool of workers who are "system-qualified." It also means a skilled man is available if parts are to be inspected on unloading when this is desirable. Users are divided about whether the crib operator's job should be part of the job rotation arrangement within a team. Manning the tool crib may or may not be a full-time job for one man.

To guarantee immediate maintenance when it is required, some maintenance personnel are generally attached to the system. This will not prevent them from helping out elsewhere, but it avoids the system manager having to negotiate for maintenance staff and it stresses the importance of the system in the eyes of the management. It is not uncommon to have a mechanical and electro-electronic technician on the main shift and one or the other on the second and third shifts.

Full descriptions of the tasks needed in running a system have been given to ensure that the jobs to be carried are appreciated. In running a system unmanned for periods, these tasks either have to be automated or have to be done in advance or at the end of the unmanned period. Tasks such as loading and unloading are difficult to automate and are more easily left to an operator. Once an operator is available to perform one task, there is every reason to exploit his presence to carry out other appropriate tasks. Additionally, the labor cost of a minimally manned system is a very small proportion of the total operating costs, and the lower manning levels become, the less incentive there is to reduce them further.

Although operating unmanned can save labor costs, it increases capital costs, as extra hardware is required. To keep the system operating during unmanned periods and during the early parts of the following manned period, a number of extra pallets and fixtures are required, the number depending on the cycle times of the parts being machined. An average pallet cycle time for many systems is of the order of 35 minutes, with the longer cycle times not exceeding 60 minutes. Thus, even if the longer-cycle operations are kept to be done during the unmanned periods, it is going to require an investment in one extra pallet and fixture per shift-hour per machine. At over $2500 each for most pallet and fixture combinations, it is relatively easy to need to find an extra $350,000 to fund the additional requirements. Pallet and fixture costs do not stop there because the loaded pallets must be stored somewhere when not being machined. On rail cart systems this may involve more plinths; on AGV systems it may involve a series of carousel storage units; and the cost of these mounts with the number of pallets requiring storage. In most instances, the work-handling-system layout will require extending to service the storage units, adding further to the cost. Additionally, because load/unloading is concentrated into the manned shifts, additional handling may be required because the load on the work-handling system is increased during the manned periods to get the pallets to and from the storage. Extra labor may also be required, as the loading requirement does not change with unmanned operation.

Tooling is another element that may involve additional costs in going unmanned. If the long cycle time operations are kept for the unmanned shifts to keep extra pallet, fixture, and storage costs low, then the tools machining the long cycle time operations will be used more frequently during these periods than would

otherwise be the case. Extra backup tools for some of the long-
cycle tooling may thus be required to be held in the ATCs. If
an operator is not present during machining, then in-process
gauging may be necessary to give the required machining accuracies, and tool-life-monitoring devices may also be required if
tools are to be used effectively. The use of these devices generally adds to machining cycles, and the devices can occupy tooling pockets on the machine. Some users find these significant
disadvantages. An alternative to tool monitoring is to use tools
conservatively during unmanned operation. But this means it is
accepted that unmanned periods operate at lower efficiency.

It can be seen that the longer the system operates unmanned,
the greater the extra costs of unmanned operation. The attrac-
tion of operating unmanned remains, however, and the instrumentation needed to support unmanned operation has been developed [5] and is being incorporated into machines. Thus, the
use of periods of unmanned operation will become increasingly
common. The economic balance between the cost of more fixtures, pallets, storage facilities, and the increased instrumentation versus the cost of an operator will, however, always need
to be investigated.

6.4 TOOL MANAGEMENT

The management and organization of tooling become increasingly
important as the variety of parts that an FMS is to machine increases. One user made a very perceptive comment when he
stated, "I had a flexible manufacturing system until I tooled it
up." Tooling here refers primarily to machining-center tooling
because users expect machining centers to provide the system
flexibility. The systems of the 1970s needed limited flexibility
in that they produced a restricted range of parts. Furthermore,
those parts had often been selected to be a family of parts, which
helped to ensure that only a restricted range of tooling was needed. The machining centers had automatic tool changers (ATCs)
which accommodated 40 or sometimes 60 tools. Users who found
their flexibility limited by this capacity developed control procedures that involved either manually handling tooling in and out
of the ATCs or accepting the fact that particular parts could only
be produced on some of the machines.

As demands on system flexibility have increased, a number of
approaches have been taken to ensure that machines have the

tools necessary. These approaches have, with one exception, required additional system hardware and control and management software. The tool management task is substantially more demanding than managing the movement of parts around a system because one part may need 30 tools to machine it and those tools need to be held on a number of machines if the system is to operate flexibly.

The one approach to tool management that avoids extra hardware and software is to control the variety of tools required to produce the parts. This requires going back to the drawing board, with the design and manufacturing departments of a company collaborating closely in rationalizing the tooling required by the parts produced by a system. It is desirable to do this anyway, but the advent of FMS has set specific targets for tool rationalization that did not exist before. It is arguably more important to control tool variety in FMS than for stand-alone machines, as a fully flexible FMS is going to require far more copies of a special tool than will be needed by an operator-supported stand-alone CNC machine. Although tool rationalization is a demanding and difficult task, it is likely to be the lowest-cost method of managing tool variety. All the other approaches that will be described require extra hardware, extra software, and extra tooling.

Increased machine flexibility can be achieved by having more tools at each machine. However, making the capacity of a drum-type tooling carousel (TC) larger than 40 tools or a chain-type TC larger than 60 tools is not very desirable because of both the drive loads necessary to power a fully loaded carousel and the time it can take to find a particular tool. The alternative approach of using two or three carousels is preferred. The Borg Warner system (Fig. 5.2) illustrates an approach to using two carousels — an active carousel and a secondary carousel. An active carousel exchanges tools with the machine spindle in the conventional manner whereas a secondary carousel swaps tools with an active carousel. Figure 6.4 shows a slightly different implementation of this approach at Normalair-Garrett Ltd. (NGL), whose FMC and integrated system was discussed in Chapter 5. The KTM machining centers, of which one is shown, have 40-tool active carousels on one side of their columns and 80-tool secondary chains of tool pockets are supported horizontally on structures beside the machines. A simple pick-and-place unit is used to exchange tools between the active carousels and the secondary carousels. Manual handling of the tools in and out of the secondary carousels is necessary in both these examples.

Fig. 6.4 View of the NGL secondary tooling carousel. (Courtesy Normalair-Garrett Ltd.)

6.4.1 Automated Tool Handling

There are a number of other ways of having more tools at a machine; these involve varying degrees of automated tool handling. Figure 6.5 shows part of an approach used by Fritz Werner, a major European supplier of FMS. In this system, a number of racks of tooling are positioned adjacent to a machine, and the tools are transferred between the racking and the machine carousel by a gantry-mounted robot. The racks are placed to the rear and alongside the machines, and the gantry can span a number of machines and a number of racks. Tools can thus be swapped between machines as well as between the racks and machines. The vertical axis on the robot enables tools to be stored horizontally in vertical racking which uses space efficiently. Fritz Werner also uses rail-cart-mounted robots to exchange tools between

Fig. 6.5 Tool rack with gantry robot in a Fritz Werner system. (Courtesy Fritz Werner/TI Rockwell Ltd.)

machines and storage racks. The rail cart runs along to the immediate rear of the machines, and the racks are positioned alongside the far side of the rails. With many machines having their tooling carousels mounted on the side of their columns, it is relatively easy to transfer tools into and out of the rear of the carousels. As with secondary carousels, the tools need manual handling into the racks. These two approaches are directly comparable with those illustrated in Figs. 5.14 and 5.13, respectively for pallet handling.

The systems just described, together with other types of double or triple TC, are all means of increasing the number of tools available at the machines. However, both the gantry-mounted robot and the rail cart could potentially be arranged to travel from the machines to a tool crib and act as a means of automatically supplying tools from a crib to the machines of a system. As the level of

flexibility needed in a system increases, it becomes appropriate
to automate all aspects of tool handling and control the organization of system tooling through a system computer, while central-
izing the preparation, maintenance, and physical management of
the tooling in a single tool crib. A number of means of doing
this have been adopted. Most have used some kind of established
work-handling equipment to get tools to and from the machines
and have then used a dedicated robot or pick-and-place unit to
exchange the tools individually with the tooling carousels. For
example, Figs. 1.1 and 5.10 show AGV layouts in which the
routes of the work-handling AGVs have been extended to include branches to the machines' tooling carousels and the tool
crib.

Rail carts do not offer the layout flexibility of AGVs, and a
separate tool-handling cart system is needed to the rear of a line
of machines if rail carts are to be used for tool handling. It is
possible to use work-handling rail carts to bring tools to the
front of any machine and use the spindle to transfer tools into
the tooling carousel. Although this is quite practicable, it
wastes time if any significant number of tools are transferred.
Mandelli, however, has used this idea rather cleverly but for a
different reason in a system they supplied to Ferrari. Some of
the Ferrari workpieces need to be machined with a long boring
bar that cannot easily be held in a carousel or handled in the
usual way. The boring bar is brought to the machines in the
fixture with the workpiece, and once machining is complete, the
spindle returns it to the fixture for return to the load/unload
area with the workpiece.

Yamazaki uses an AGV to transport tools from the tool crib
to the machines but uses the approach illustrated in Fig. 6.6 in
some of their systems. Tool carousels act as tooling pallets, and
these are brought and exchanged at the machine rather than individual tools. A carousel may also be stored at a machine (as
shown to the left of Fig. 6.6), and this acts as a secondary carousel that is swapped with the active carousel rather than having
single tools exchanged. This approach has features of the original Molins approach, in which a workpiece and the tools to machine it were both brought to a machine.

One of the largest automated tool-handling systems in the
world is at the Augsberg, West Germany, plant of Messerschmitt
Bolkow Blohm (MBB). MBB machines and assembles parts of the
Tornado and European Airbuses at its Augsberg plant [6], and
to ensure speed and reliability in all stages of this work, MBB

Fig. 6.6 Yamazaki changeable tooling carousels. (Courtesy Yamazaki Machinery UK Ltd.)

has invested heavily in high technology. Within the plant it has a machining facility termed the computerised integrated automated manufacturing (CIAM) system. This consists of 24 machines under DNC control. There are 11 portal milling machines, five horizontal and four vertical milling machines, and four machining centers. The machines have free-standing tool magazines, which are linked to the tool storage and maintenance area by an overhead monorail. The monorail delivers preset tools to the machines in pallets from which they are loaded into the machine or to the magazines. The scale of this system is impressive, not only because it caters to the needs of 24 machines but because of the volume of tools that are handled and managed. The CIAM system is just one of the automated facilities at the Augsberg plant.

All the approaches described are designed to ensure that system flexibility is not limited by not having the necessary tools at

machines for a part to be machined. It will be appreciated that
the scale of the control system and extra software needed to man-
age the tool scheduling and transporting is substantial.

6.5 CONTROLLING PRECISION

When the parts to be produced on an FMS have close tolerances,
special steps may be taken to achieve the required accuracies and
to check that they are being achieved. This may include the selection of particular machines (as was discussed in Chapter 4),
and it may also involve the inclusion of one or more coordinate
measuring machines (CMM), which will be served by the system
work-handling system. Figure 5.5 illustrates a CMM that is part
of the Hughes Aircraft FFS. The systems whose layouts are
shown in Figs. 1.1 and 5.10 also include CMMs. CMMs are CNC-
controlled and are programmed to inspect a part and record all
the readings taken.

Parts may also be inspected by machine-tool-mounted probes.
The most common form of these is the spindle-mounted probe on
machining centers, which is carried on a standard tool taper and
held in the tool carousel when not in use. Such probes can be
used to measure the position of a fixture or part datum before a
part is machined, and the worktable can then be offset by the appropriate amount. This eliminates any errors arising in locating
the part in the fixture or the fixture on the pallet or the pallet
on the machine. It is thus a very beneficial form of measurement
in that it reduces some potential machining errors before they
occur.

The main reason users incorporate a CMM into a system is to
help control quality rather than just to inspect finished parts.
The control of quality implies using the inspection data positively,
as in the probe example. Thus, if a bore is checked and found
to be undersized, then this indicates with a high degree of certainty that the tool that produced it has either worn or was preset inaccurately. If some means exists of automatically adjusting
the tool, then this can be done, or (more likely), the tool is selected to be returned to the tool crib for resetting and an identical tool, termed a sister tool, is selected next time the tool is
required. To do this requires the system software to store his-
torical usage data about every tool because by the time a part
has completed being inspected on a CMM, the machines and tools
that machined it will be engaged in machining another part. If

the faulty boring tool was near the end of its programmed life, it may even already be on its way to the tool crib. Thus, the extra system software required to use CMM inspection data effectively is quite complex, as it has to keep track of all the tools used, it has to relate their usage to the machining of particular features of particular workpieces, and it has to contain algorithms to make logical decisions based on the results of comparing inspection data with toleranced reference dimensions for a workpiece.

Larger bores on parts may be produced by circumferential milling rather than by use of a boring tool. If such a bore is found to be undersized, then it may again be concluded that tool wear has taken place, and again one means of overcoming this is to return the tool to the tool crib. Alternatively, although the tips of a milling cutter cannot be adjusted automatically on a machine, the path it cuts can be adjusted by altering the path it is instructed to follow, i.e., a correction through software. The use of tool offsets is common enough on CNC machines, but these only work on one axis. A diametral offset requires a program code to be modified, not just an offset changed. The structure of some machining programs facilitates such modifications, but the question arises, should a program be modified to accommodate a particular tool and a special version of a program stored and used until that tool is returned to the tool crib?

Correcting undersized bores is simpler than correcting most other types of error because the causes of other errors are not as easily identified. For example, if the relative position of two bores is checked and found to be outside specification, what has caused the error? Has the machine distorted slightly, the coolant temperature control failed, or what? Thermal distortions in machines pose the greatest difficulties in achieving high accuracies on machines, as the degree of distortion depends on the duty cycle of the machine and this depends on the parts it has cut and the sequence in which they were cut. Manufacturers' appreciation of thermal distortion problems of machine tools is now greater than it has ever been, and design practices are followed to minimize the effect of heat sources. Temperature-controlled coolant is also often used to maintain the workpiece at a steady temperature, and some systems are installed in temperature-controlled machine shops to help keep the machines at a steady temperature. However, thermal distortions are still almost bound to occur.

Studies are being carried out (for example Ref. 7) to create thermal models of machines so that any distortion can be related to possible errors arising in a workpiece. Once the relationships are established from machine to workpiece, it will be possible to write software that responds to readings from thermocouples mounted on machines and compensates the machining program for potential errors. This will require, however, further instrumentation on machines and considerably increased software sophistication.

This review of controlling precision in FMS is, of necessity, only a brief introduction to some of the problems and some possible solutions. Fortunately, only a few workpieces require machining to tight tolerances and those that do generally have a limited number of critical features. Because of this, the most effective way of exploiting a CMM in an FMS is to use it to carry out process capabilities studies. The critical features of any workpiece can be identified and plans made to control these features while ignoring the rest. The control strategy adopted will then need to be incorporated into the system control software.

6.6 THE DEVELOPING FMS SCENE

The aim of flexible manufacturing can be summarized as to provide a capability to respond to market demands with minimum lead times and with cost-effective utilization of floor space, equipment, and personnel. The existence of actual systems and the interest that has been shown in them have helped to highlight these aims, and they are now accepted as desirable for all manufacturing systems. Their value has also been appreciated in a wider context such that they have also been adopted as company objectives by many companies. This general relevance of the aims of flexible manufacturing has been emphasized in this book by referring to flexible manufacturing as a philosophy of manufacture, while referring to FMS as manufacturing systems of various types through which that philosophy is implemented.

In Chapters 2 and 3, it was explained how the flexibility of the horizontal machining center facilitated the development of the original metal-cutting FMS. Those looking to extend the areas of application of FMS have looked for similar flexible machines or processes to form the building blocks of FMS, because flexible manufacturing is capable of application to all areas of manufacture where there is operational flexibility in a process or a machine. If a machine can be readily reprogrammed and automatically reset to process a different part, then it can form a building block of an FMS. This has been realized by the designers of many types of machine tools and by the designers of other types of manufacturing system, and hence developments are taking place on a broad front to increase machine and process flexibility so that the benefits of flexible manufacturing can be exploited. The following sections give examples of recent developments that either are aimed at or have succeeded in extending the flexible manufacturing approach. The first of these concerns increasing the flexibility of lathes.

6.6.1 Shortening Lathe Setup Times

In Chapter 4, the difficulties of resetting lathes automatically were discussed, particularly in relation to changing chucks. It was shown that this meant that fully flexible operation in terms of machining a variety of turned parts in any sequence was generally not practicable, although developments are in hand to make this practicable. The other element of setup on a lathe relates to tooling. For many years, a few of the larger lathes have been designed with automatic tool changers. Like the ATCs of machining centers, these have changed not only the tool, but the complete tool assembly as well. The approach has not been widely adopted because it has been found more cost effective to increase the number of tools held per turret and to increase the number of turrets. Increasing the number of tools on a lathe both increases and decreases its flexibility. Flexibility is increased because of the number of tools available to perform different cutting operations. Flexibility is decreased because of the additional time needed to change the tools on setting up a part that cannot use the existing tools. To help shorten the tool-changing time of turret-based tools, a number of manufacturers have developed schemes to ease (and in some cases automate) changing tools [8].

Figure 6.7 illustrates the main elements of one approach to this developed by Sandvik Coromant. The system is called "block tooling." As may be seen in Fig. 6.7a), the conventional shank of a lathe tool is shortened to a small block that is clamped onto appropriately modified turret faces of a lathe by a spring-loaded draw bar. The geometry of the interface ensures repeatable location of the blocks, while providing adequate clearance for the blocks to be slid on and off easily. Block changing takes place only at one turret position, which has a hydraulically actuated plunger to release the drawbars. Blocks can be changed quickly and easily either manually or automatically by using a gantry-mounted changer unit that accesses the blocks required from racks of a magazine situated at the end of the machine, as shown

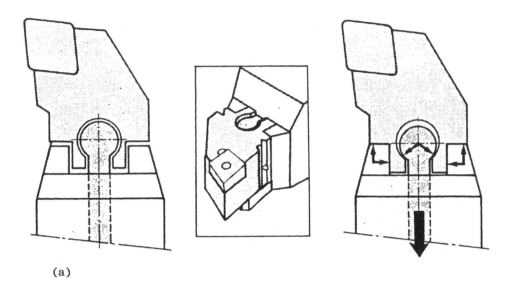

(a)

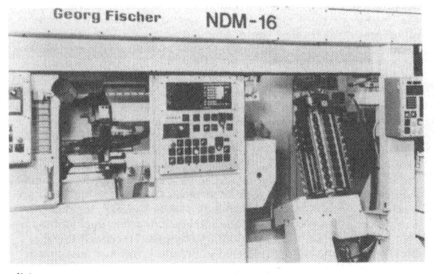

(b)

Fig. 6.7 Aspects of "block tooling." (a) Interface design. (b) A lathe with block-tooling racks in a magazine. (Courtesy Sandvik Coromant.)

in Fig. 6.7b. The block tools need to be measured before being inserted into their turret holders to determine the tool offsets from their nominal positions. These figures must then be entered into the CNC controllers. With automatically changed blocks it is more usual to have a data link between the measuring unit and the CNC controller so that the offsets are automatically transferred.

6.6.2 Sheet Metal FMS

Sheet metal parts have traditionally been produced by a range of shearing, forming, and forging machines, but the last 15—20 years has probably seen more significant changes in the technology of these machines than in that of metal-cutting machines. The technological changes include CNC, multioperation machines, and laser and plasma cutting.

Just as machining centers have made many individual drilling, milling, and boring machines obsolete, high-speed sheet metal machining centers (SMMC) that can punch and nibble have replaced individual punches, shears, presses, and nibblers for producing parts from steel sheets and plates. This development has involved imaginative machine design and has been possible partly because of CNC and partly because of the automatic tool/ die-changing facilities fitted to these machines. Thus, as with machining centers, many sheet metal operations can now be carried out at one setup under programmed control, and tool changing is quick, easy, and automatic. A flexible machine is thus available to form the basis of a flexible system.

Not very long ago, there was a clear distinction between sheet metal parts and parts made from plate. The boundary between sheet and plate occurred at about 6 mm thickness (1/4 in.). Parts below that thickness were sheared, cut, punched, nibbled, and bent on press brakes by sheet metal workers. Parts over this thickness were flame-cut, rolled, and welded to shape by boilermakers. The new sheet metal machining centers have more localized power than their predecessors, thus enabling them to handle sheets up to 13 mm (1/2 in.), and sometimes thicker. This has helped to blur the boundary between sheet and plate. The development of laser and plasma cutting, which are capable of handling even greater thicknesses, has helped to distort the boundary further because some machines now combine laser and/ or plasma cutting with punching and nibbling operations. The flexibility of the machines is thus increased further. The versatility of some of these SMMC is now such that they are almost manufacturing systems on their own.

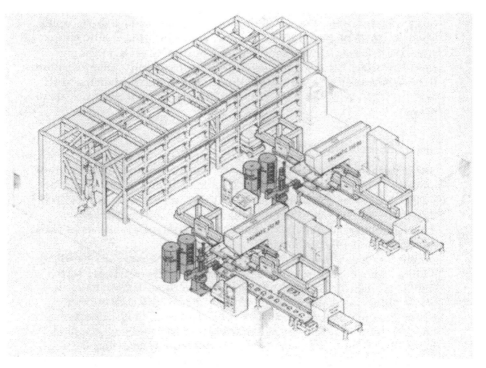

Fig. 6.8 A flexible manufacturing system for sheet metal parts. (Courtesy Trumpf Machine Tools Ltd.)

Figure 6.8 illustrates a flexible system that integrates two such machines and an AS/RS with stacker crane. The stacker crane does not leave the AS/RS, but transfers sheets from the internal racking to simple rail carts which come out from the sides of the AS/RS to supply the machines. (A view of the central aisle of the AS/RS is shown in Fig. 5.12.) The two machines have automatic loading and unloading transporters, and each machine also has two carousel storage towers of the die sets. These are accessed by simple robots to supply the machines with the necessary tools so that jobs can be processed in any order. The only ancillary operation handled by the system is deburring, with one deburrer being situated on the output conveyor from each machine. Sheet metal flexible systems generally take the form of unidirectional link lines with a punching/nibbling machine or

SMMC and perhaps an AS/RS at the head of the line. The speed of operation of forming machines is quite high, so that a link line of a few individual machines is often adequate to meet a company's requirements without having identical operations carried out in parallel, as in metal-cutting FMS.

The late 1980s are gradually seeing further examples of such systems being installed. A recent example of such an FMS is the system installed at Xerox's Component Manufacturing Operations plant in Webster (NY) to produce side panels for copiers [9]. This system has a programmable punch, which accepts sheared blanks linked to a programmable bending press, which in turn feeds a plasma arc welder. These three machines are linked by a work-handling system that includes a blank turnover unit prior to the bending machine. The panels are then manually loaded onto robotic spot welders. The system will initially produce 10 different panels which were previously produced by a series of single-function machines throughout the plant. The flexibility of the punch is such that it has 23 blanking punches, three auxiliary forming punches, and a right-angled shear.

Sheet metal FMS have a great advantage over metal-cutting FMS in having little variation in the product they handle. It is all sheet. The sizes and thicknesses of the sheets may vary, but it is relatively immaterial whether small parts or large parts are produced. The flatness of sheets means that simple conveyors or suction devices are adequate for handling, and a range of sheet sizes can be handled with similar equipment. Flexible handling equipment is not required, and machine loading and unloading can be relatively easily automated. The costs and complications of using pallets and fixtures are avoided.

6.6.3 Flexible Assembly Systems

Assembly has not been unaffected by the increased awareness of the philosophy and benefits of flexible manufacture. Many have rightly assumed that assembly is already a type of flexible manufacture because the majority of assembly is carried out manually and a human operator, with a selection of powered tools, is one of the most flexible "systems" there is. However, in this section purely manual assembly systems will be ignored and examples will be given of systems where combinations of human operators and automation form flexible assembly systems. This combination of human operatives and automated machines and/or handling systems is, of course, also found in metal-cutting FMS.

Machines for automatic assembly have existed for a long time, but these have mostly been fixed-automation machines. It was the advent of the industrial robot with its ability to be programmed, and hence operate flexibly, that spurred the development of flexible assembly. In the 1960s and early 1970s, industrial robots were applied to operations (such as spray painting) and handling tasks (such as unloading furnaces) where the environment was unpleasant and often injurious to human operatives. These tasks did not make high demands on the positioning accuracy of the robots. The early robots were quite large, and this meant they were difficult to control to the accuracies necessary to assemble close-toleranced parts and thus their use for assembly was not too seriously considered. However, as the design and performance of robots improved and their use for welding car body panels together became established (fabrication is, after all, an assembly operation), their application to other types of assembly started to receive much more thought. The accuracy problem was eventually overcome by changes to both the robots and the products that were to be assembled by robots. From the robot end, smaller robot arms (such as Unimation's PUMA, introduced in 1978) were brought onto the market especially designed for performing assembly tasks. The limited reach and small size of these arms both helped them have improved accuracy and enabled them to be positioned over or beside assembly conveyors without being a danger to anyone working nearby. From the product end, products were redesigned to ensure that assembly by robot was both practicable and easy.

Because robots have CNC controllers, they can easily store a number of programs to vary the tasks they have to perform. Programmed instructions can include instructions to change the grippers or the tools they are using in addition to changing products [10]. As with the families of parts machined by metal-cutting FMS, a robot will typically assemble, or contribute to the assembly of, a family of products. This is necessary because the flexibility of a robot is limited by the flexibility of the grippers and the tooling it has available. The applications engineering necessary to use a robot can be considerable. It has two elements, the design of the workstations and the design of the grippers and tooling attached to the robot. The workstation has to be designed to position the base unit to which parts are to be added and to hold suitable magazines of the parts to be assembled so that these can be accessed by the robot. Space also has to be made for the various grippers and tooling needed by the robot. Economics and space limitations dictate that the range of grippers and tooling used and the number of parts held at a workstation must be restricted. Hence robots tend to be applied to medium- to high-volume production applications of a restricted product family.

A ready example of robot-based flexible assembly is the mounting of electronic components on printed circuit boards (pcbs). The boards themselves are a standard size (giving a product family), but various types of component are mounted on them in a variety of configurations depending on the design of the pcb. This is a good application first because the components to be handled are very small and second because their specification may vary while their size remains the same (resistors and capacitors are examples). Thus, a few grippers can handle a range of components. An assembly robot (or sometimes a programmable assemble machine) will generally be able to handle most of the components, but there are invariably some that are specialized. Rather than design special grippers to handle these, it is more usual for them to be manually assembled. This combination of manual assembly with a flexible machine or system is typical of much of today's flexible automated assembly technology.

A further example but different in both scale and detail is provided by the Leyland Assembly Plant at Leyland, England. Leyland produces trucks in a number of model families but with many variants on each family to suit customer requirements. The complete range of trucks produced by the company is assembled on just two assembly lines, one for larger trucks and one for smaller. The trucks are assembled as one-offs to a particular build schedule so that every truck on the line is invariably different from its neighbor — a truly flexible assembly line. The keys to this approach being practicable are the use of a computer-controlled materials management system, which ensures that the parts for a particular truck are available to be assembled when needed at the lineside, and the flexibility of the human operatives performing the assembly tasks.

The final example of a flexible assembly system that will be described here integrates assembly and testing and an element of JIT manufacturing. It also uses a stacker crane and an AS/RS in a novel way. The system, known as Project Mercury, was installed at the International Computers Ltd. (ICL) plant at Ashton-under-Lyne, near Manchester, England, in 1985, having been supplied by Fenner Systems Engineering following conceptual design by

ICL engineers. The system was conceived as a flexible assembly
and testing facility for the ICL Series 39, Level 30 range of computers and other computer products and systems with cabinet
sizes approaching 1 m^3 (the product family). The Series 39 consists of five basic cabinet types, but each cabinet can have many
configurations and different internal components depending on
customer requirements. Each cabinet thus has different assembly and testing procedures. The problem is identical to that of
the Leyland Vehicles example, but the products and procedures
are very different.

The hardware of the system comprises two parallel, multitiered,
70-m (230-ft) long rows of racking which are served internally by
a stacker crane. (It is misleading to call this an AR/RS for reasons that will become clear. The stacker crane is simply the work-
handling unit of the FMS.) Externally, the lower tiers of racking
have metal roller doors which allow the racking locations to be accessed by either assemblers or testers. One row of racking and
the area beside it is used to progressively assemble cabinets (see
Fig. 6.9); the other row holds finished cabinets and computer
systems (comprising a number of cabinets) for a series of testing procedures to be carried out (see Fig. 6.10). Computer cabinets are assembled in stages as they move from one end of the
racking to the other down the assembly side of the system and
are then progressively tested as they return down the other side.
Input and output occur via a conveyor through six racking locations at one end of the racking.

For assembly to take place at an assembly station, initially a
base unit and subsequently partially assembled cabinets are withdrawn from the racking onto floor-mounted supports, as illustrated in Fig. 6.9. The racking locations adjacent to the assembly stations hold parts containers which hold trays of the parts
necessary for the assembly operation. The assembler has the
necessary instructions for the assembly operation displayed on
a VDT (VDU). Once the assembly operations for a particular
product at a particular station are completed, the assembler returns the cabinet to the racking and closes the door, thus indicating to the control system that the operation is complete. The
control system then sends the necessary signals to the stacker
crane to move the cabinet and the parts containers if appropriate.
The cabinet may be moved to its next assembly operation or it
may be temporarily stored. In-process storage is provided by
the top tiers of racking.

The testing side of the system is similar to the assembly side
except that two levels of racking are used for holding cabinets

Fig. 6.9 Project Mercury: An assembly station showing adjacent parts containers in the racking. (Courtesy Fenner Systems Engineering.)

(as shown in Fig. 6.10) and the completed computers/cabinets are generally not withdrawn from the racking. The stacker crane turns the cabinets so that their backs are accessible for testers to make connections to nearby testing computers and to a main supply. The system has the capacity to assemble and test 15,000 cabinets per year.

As with all FMS, the flexibility and operational efficiency of this system is very much determined by the sophistication of the scheduling and control software. The system has four levels of control, resembling those shown in Fig. 6.2. The lowest level, the device level, has various computers controlling the crane, input/output conveyor, and workstations. Level 2 is used for process control. Level 3, based on an ICL System 25, looks after operational planning and control, while strategic planning takes place on an ICL 2988/3980 at level 4. Mention was made

Fig. 6.10 Project Mercury: The testing side of the system showing the cabinets and adjacent space for all the testing equipment.
(Courtesy Fenner Systems Engineering.)

earlier of the system incorporating a JIT philosophy. The JIT
philosophy is incorporated into the control software by having
a customer order start a chain of pulling a product through the
system, as described in Section 3.4. JIT is partly physically implemented through both the parts containers and pallets on which
the cabinets are built. These are coded, and they act as Kanbans
when sent to the main parts stores and thus start a build cycle for
a particular computer model.

6.7 REVIEW

The flexible assembly and flexible sheet metal manufacturing systems that have been described illustrate that flexible manufacturing

is now accepted as applicable to much more than just machining systems. However, the current growth in the number of machining FMS means that these will still be the most common type of FMS for some time to come. Most of these will be what might now almost be called the traditional type, with work handling provided by wheel-based vehicles linking a number of machines. Others will doubtless break new ground, as the forklift AGV did when it opened up a third dimension to materials handling in FMS. Greater integration is likely in terms of what is within the system (such as AS/RS) and by linking systems using the developing protocols.

The various types of system described all illustrate that the technology of FMS is advancing. One consequence of this is that the distinction between what is and what is not a flexible system will start to blur. This is already happening at the FMS:transfer line boundary. Some transfer machines incorporate head changers or head indexers, giving them some flexibility. The part family that the transfer machine can accept may only be two or three parts, but it is no longer a completely dedicated machine. Other transfer machines have had their flexibility increased further by having machining centers complete with tooling carousels mounted at transfer line workstations and by having quick-change heads on the head indexers used in the line [11]. The head indexers have been designed so that a complete headring can be changed in under 2 minutes. The result of changing a headring is that the transfer line now machines a different family of parts. Should this type of line also be termed a flexible transfer line or should the batching feature of different setups be somehow recognized in its title? The jargon has not yet caught up with the technology.

Parallel developments are occurring with some machining centers. Some HMCs are now being designed so that they can machine with and automatically change small unit heads. The idea of this is to improve the productivity of HMCs when drilling hole patterns. This borrowing of ideas and hardware across systems can be expected to grow. This means that the classification of manufacturing systems as given in Fig. 1.2 needs to be modified to reflect the realities of the 1980s and beyond in terms of the divisions shown in Fig. 6.11. At the mass production end, there will probably always be a place for fixed automation transfer lines (TL), although the number of applications is likely to reduce and these be taken over by flexible lines (FTL). Flexible manufacturing started to be implemented at the high-volume end of batch manufacturing, but more recent systems have progressively extended its range of application to increased variety and smaller

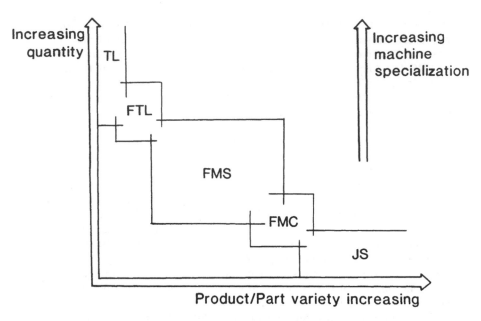

Fig. 6.11 A structure of the modern manufacturing scene. TL, transfer line; FTL, flexible transfer line; FMS, flexible manufacturing system; FMC, flexible manufacturing cell; JS, job shop.

volumes. It is likely that batch manufacturing will increasingly be taken over by FMS, FMC, and just-in-time (JIT) manufacturing philosophies.

In Chapter 3, it was explained that flexible manufacturing includes most of the ideas of JIT, as do TL and FTL. The element that is missing is the ideas of parts being produced on demand from their next operation rather than being pushed through as the result of an order to start manufacture. This topic is part of the wider subject of manufacturing planning and control, which includes materials requirements planning (MRP), manufacturing resource planning (MRP II), and computerized manufacturing scheduling. The other significant technology required to be able to produce almost any part or product on demand (or at least with very short lead times) is the use of CADCAM in the design and manufacturing planning offices [12]. These topics are beyond the scope of this book although they are very relevant to the environment in which flexible manufacturing operates. A company needs a balanced manufacturing strategy in all these areas if it is to maximize the benefits it will gain by investing in any one of them.

This chapter concludes the section of the book that has dealt with the development of flexible manufacturing and its hardware and systems. The remainder of the book presents and explains the software tools available for designing and analyzing FMS and related manufacturing systems.

REFERENCES

1. J. C. Albus, C. R. McLean, A. J. Banbera, and M. L. Fitzgerald, "An architecture for real-time sensory-interactive control of robots in a manufacturing facility." 4th IFAC/IFIP Symposium on Information Control Problems in Manufacturing Technology, Gaithersburg, October (1982).

2. Anon., *MAP, Manufacturing Automation Protocol*, General Motors Technical Center, Warren, MI (1986).

3. P. Evans, "MAP and CIM," *Data Processing, 28*, 3, 151–156, April (1986).

4. H. Yoshikawa, "The Japanese project on the automated factory." 3rd IFIC-IFAC Int. Conf., "Prolamat 76," Stirling (UK), June (1976).

5. J. Tlusty and G. C. Andrews, "A critical review of sensors for unmanned machining," *Annals of CIRP, 32/2*, 5, 563–572 (1983).

6. G. Handke, "Computer integrated and automated manufacturing systems in aircraft manufacturing -- User's viewpoint." Proc. of 4th Int. Conf. on FMS, Lindholm, R. (ed.), Stockholm, IFS (Publications) Ltd., Bedford, UK, pp. 533–557 (1985).

7. R. Venugopal and M. M. Barash, "Thermal effects on the accuracy of numerically controlled machine tools." Final Report, Vol. 4, ONR Contract No. 83K0385, Purdue Univ., West Lafayette, IN, October (1985).

8. J. Scherer, "Tool and work changers for metal cutting machine tools," *Industrial & Production Engineering, 10*, 1, 14–19 (1986).

9. J. A. Vaccari, "FMS blanks, forms, welds steel," *American Machinist, December*, 71 (1985).

10. T. J. Drozda, "Flexible assembly system features automatic set-up," *Manufacturing Engineering*, December, 75-76 (1984).

11. A. C. Montag, "Flexible automation for high-volume production," *Manufacturing Engineering*, November, 79 (1984).

12. M. D. Groover and E. W. Zimmers, Jr., *CAD/CAM: Computer-Aided Design and Manufacturing*, Prentice-Hall, Englewood Cliffs, NJ (1984).

7

Simulation and Analysis in the Design of FMS

7.1 INTRODUCTION

Computer-aided-design (CAD) software and hardware may be used to assist a designer by easing the transition from the designer's concept of product to the process plans for making the product. A grander concept of CAD (called CADM, say) would carry the transition past the process plan stage and further consider the impact of product manufacture on design (as suggested in Fig. 7.1). The software for CADM would incorporate models of the manufacturing process as well as models of the process planning activities. Here, a "model" is any (partial) representation of reality. Most often, we will be interested in those representations which can be implemented as a computer program. Running such a program presumably provides information to the designer that could not easily be obtained in any other way. (For example, it is seldom feasible to construct a manufacturing system in order to assess its impact on the product design. It is usually easier and less expensive to analyze a model of it.) A grouping of models that represent the process planning and manufacturing systems that impact on product design has been referred to as a "virtual manufacturing

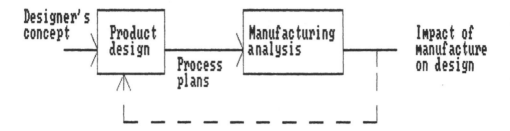

Fig. 7.1 CADM procedure.

system" [1]. The models of manufacturing systems discussed in
this portion of the book could be incorporated into a CADM system. Our purpose here, though, is more particularly to use the
manufacturing system models developed over the past decade to
aid the FMS designer who may be called on to design an FMS or
analyze the FMS's behavior.

The basic philosophy of designing and operating an FMS, by
now well established, is the systems approach. Hall [2] presented a morphology for systems engineering. Gibson [3] proposed a five-step procedure for systems engineering which Sage
[4] noted is equivalent to Hall's. The five steps can be summarized as follows:

1. Development of goals
2. Establishment of criteria on which goal achievement can be
    judged
3. Development of alternate candidate solutions
4. Ranking of alternatives by applying the criteria to the alter-
    nate solutions
5. Iteration of the above four steps to obtain a deeper analysis
    of alternate solutions and to converge on an acceptable solu-
    tion

The initial step of developing goals includes a statement of
what ought to be. This is compared to what is (i.e., the current situation). The comparison results in the definition of a
set of needs that the new system should satisfy. To determine
whether the system meets its goals (i.e., satisfies its needs),
measurable criteria are established which presumably correlate
with the goals. For example, the performance goals of an FMS
might be measured by production rate, machine utilization,

throughput time, and so forth. (In fact, the emphasis in the remainder of this book is on the means by which a designer can quantify these measurable criteria without incurring the expense of actually constructing the system.) Alternate designs for an FMS may then be compared with regard to these measurable criteria. If none of the alternative designs is satisfactory, then the designer may repeat some or all of the steps of the system engineering process. This iterative process will be referred to as the system engineering design process.

7.2 THE DESIGNER'S DILEMMA

It is common knowledge today that the need for an FMS is most acute for those products or part types that are produced in intermediate volume (say on the order of 10,000 parts per year). Moreover, these part types should share some similar work content, and there should be a reasonable expectation that they will be needed for some years to come. Determination of the degree of similar work content is aided by having taken a group technology approach to manufactured parts. (Conversely, attempts by a company to develop an FMS system for its needs will probably result in more serious consideration being given to a group technology study of parts manufactured by the company.) The requirement of continued need for the produced part types is a little less critical for FMS than for conventional facilities. This should be so owing to the inherent flexibility of the FMS, which can theoretically allow for production of a similar group technology class with only moderate modifications to the FMS.

A company may designate, perhaps via group technology, one or more groups of part types as having similar processing. Given one such group, the work content of those part types begins to define the requirements for the FMS that will produce them. However, since even the part types within one group will have possibly very different value to the company, it may not be desirable to produce all of the part types within the group. In that case, the choice must be made as to which part types will be produced on the FMS. Furthermore, it may be the case that the size of this chosen group is simply too large, and not all members of the group can be made on the FMS at the same time. The choice of which part types to produce on an FMS during a specified time period clearly depends on the kinds of processing that the FMS can perform. But in the design phase of an FMS, this results in a circular kind of "catch-22" reasoning; i.e., the FMS

cannot be designed until the part types are selected, and yet the
part types cannot be selected until we know the FMS on which
they are to be made!   There are at least two approaches to resolving this dilemma.

Both approaches to resolving our designer's dilemma involve
the basic issue discussed in this portion of the book. That issue is the use of modeling to aid the design process. For example, modeling will aid the system engineering design process
by providing quantified values for the measurable criteria used
to compare design alternatives.

7.3 MODELING APPROACHES TO SOLVE
THE DESIGNER'S DILEMMA

The explosive growth in the number and types of FMS models
can be traced in part to a National Science Foundation (NSF)
program in Production Research begun in the mid-1970s. By
1974, a few FMS were installed and showing need for improved design and operating procedures. These early systems became the
focus for research and development of modeling tools that could
potentially aid in that improvement. One of the NSF grants at
Purdue University resulted in the development of several tools
based on either computer simulation or queueing-network theory
[5] (both these tools are discussed extensively in later chapters).
Research at C. S. Draper Laboratory (Cambridge, MA) provided
for further development and proliferation of modeling tools to aid
the FMS designer [6]. The Draper group had as one of its objectives the solution of what we have referred to as the designer's dilemma.

One way to handle the designer's dilemma is to use something
like the system engineering design process. The goal in this
case might be to produce all the part types in the group technology (gt) class on one FMS. The approach could be to select
an initial subset of part types (based on management intuition)
and then design an FMS for those part types. A candidate FMS
design may be arrived at, for example, by the methods and
models discussed in the next four chapters. Its performance criteria can then be evaluated for the addition of further part types
in the same gt class. If performance degrades, the FMS design
itself may be modified to restore lost performance. Then, if
successful so far, further additions of part types may be made
until hopefully the entire gt class can be produced by the FMS.

Whether and when this iterative process converges to a satisfactory answer depends, in part, on the ingenuity of the designer in finding modifications that restore performance.

A second approach to our designer's dilemma is a more global one. That is, the problem is stated in its entirety and solved without iterative design. The framework in which the problem is stated is that of mathematical programming. In this context, all part types in the group are considered at once, as well as all machines that are candidates for the FMS! A measure of performance is specified that deals with the cost of the machines finally selected, as well as with the value of the part types finally selected for production on the FMS. The idea is to minimize the cost of those machines while producing the most valuable part types on the FMS. Chapter 12 contains a description of this approach to the machine-part-selection decision.

The iterative approach to the FMS design problem mentioned above called for an evaluation of FMS behavior or performance even though no physical system existed on which to measure that performance. We shall show how that performance can be estimated using models that portray the dynamic behavior of parts, machines, tools, and materials handling in the FMS. Two kinds of models are used to consider the dynamic interactions in the FMS: computer simulation models and network-of-queue (n-o-q) models. Both types of models represent any FMS as a set of processing stations where parts may wait for processing, and where parts may move among the stations in a wide variety of patterns. These model types are both essentially descriptive in nature. That is, the output of either type of model only describes the behavior of the system in that situation, but does not prescribe what action to take given a set of manufacturing situations. For example, a descriptive model would describe the behavior of an FMS with five AGVs for material handling, but it would not tell the designer how many AGVs should be in the FMS for highest production rate. Thus, these types of modeling seem not to be directly applicable to synthesis activities. However, the designer can use these models for design purposes in an iterative way, in the manner described for the system engineering design process.

We can restate that design process for our view of FMS design (as illustrated in Fig. 7.2). First, the designer uses intuition, experience, and rough-cut models (described later) to arrive at a tentative or candidate design. Then the performance criteria values for this design are obtained via n-o-q or simulation modeling. By studying these criteria values, the designer

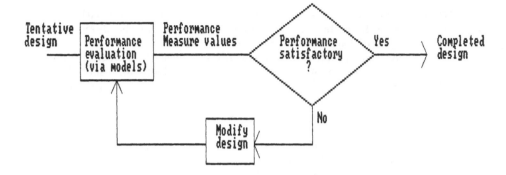

Fig. 7.2 FMS design as a system engineering design process.

can determine whether the design is satisfactory and, if not,
possibly identify aspects of the design that, if changed, might
result in performance improvement. The corresponding changes
are made in the model, and the performance is again estimated.
This iterative process continues until the designer is satisfied.

As shown in the next four chapters, the two types of descriptive models mentioned above are in fact complementary to each
other in that the simulation models are in effect more detailed
models of the system than depicted by n-o-q models. Because
of this difference in model detail, it is more appropriate to use
n-o-q models in the early stages of design. Simulation models
can be based on the resulting n-o-q models, and they usually
find better utility after the system design has reached an advanced stage, or in fact when the system goes into production.
This is because congestion in the system, overflow of waiting
space, differing prioritization of workpieces, and so forth, are
detailed behaviors of an FMS that have significant impact on the
system operation, but may be ignored at the early design stages.

7.4 MODELS CAN ASSESS THE NEED
FOR RESOURCES

So far, we have contended that the parts to be made on an FMS
affect the design of the FMS (and vice versa). If we are given
the parts to be made and the machines to make them, we still
need to lay the machines out in some physical configuration

and provide for handling of parts (and, for some systems, tools)
in a computer-directed control scheme. The machines, tools,
fixtures, and material-handling devices may be referred to as
the resources of the system which can facilitate the production
performance if provided in proper quantity and type. Concern
with number and type of machines to be incorporated into the
FMS is appropriate at early stages of FMS design. Information
to respond to those concerns is obtained from n-o-q modeling.
As design progresses, emphasis is placed on specifying physical
distances between machines and the speed and efficiency with
which these distances may be traversed by material-handling
equipment. Also, there may be many ways in which parts (and
tools) can be moved among stations, some of which can result in
significant traffic congestion. To model the dynamic behavior of
the FMS at this level of detail, a computer simulation is usually
employed. As with all stages of design, this one is iterative and
may require the designer to run and rerun the model many times
before a satisfactory result is achieved. In Chapter 11, we shall
consider the potential for establishing an automated way of using
simulation models in order to relieve the tedium and frustration
that might be experienced by the human designer.

By definition, an FMS allows for some flexibility in its operation. Such flexibility results from incorporating in the system
machines that can provide a fairly wide variety of operations.
To do so, these machines need, in turn, to employ a variety of
tools with which to do their work. The tools must either be stored
and accessed at the machine or be delivered quickly to the machine on demand. These two types of tool management represent
widely different strategies of resource management. For the case
of tools stored at the machine, we can envision two or more copies
of at least several types of tools to be placed at different machines
(for reliability in case of breakdown and to relieve potential bottleneck situations). If large numbers of redundant tools are stored
at the machines, the resultant cost of tooling alone could be significant. Managing the tools implies that a batch or mix of part
types is specified, and that tools to perform the required processing must be assigned to machines in an efficient and effective manner. This assignment problem is called "loading" the FMS. In
Chapter 12, we shall consider both exact (i.e., mathematical programming) and heuristic methods that have been developed for solution of the loading problem.

The second type of tool management strategy involving tool delivery is appropriate where potentially redundant tools are quite

expensive. Perhaps more significantly, management strategy for
this tool-delivery FMS need not be as concerned with the very
difficult loading problem. For the case of ever-changing part
mixes, there will be a continual need to solve the loading problem for non-tool-delivery FMS. Tool-delivery systems can in a
sense be adaptive to the changes in part mix (and perhaps tool
reliability) that occur at the FMS. Such adaptability is obtained
at the price of more complex hardware and software. This strategy has been studied from the standpoint of simulation and n-o-q
models. In these cases, as would be expected from inherently
descriptive models, the results are derived only for specific types
of FMS configurations. The models used for those studies are described in Chapter 11.

The design of software for control of the FMS includes computer control of each machine as well as control of the material-
handling system (mhs) between processes. The control of each
machine is usually governed by software that is outside the domain of the software tools considered in this book. We can, however, be concerned with the design of computer control of the
mhs devices. The software tool of choice for this kind of design
task is simulation. Analytical tools (such as n-o-q models) by
and large lack the precision of detail needed to represent the material movement. The computer software for control of material
handling can be complex and can strongly affect the effectiveness
of the entire FMS. This software must account for possible traffic congestion in the system and may also be called upon for relief
in special situations like breakdowns or the discovery of poor
product quality. No general design procedure for the mhs control software is available at present. Simulation aids this design
process in its usual way, namely by describing the behavior of
design alternatives and providing information on those alternatives to the designer. The designer can then develop a new design and proceed in an iterative manner.

7.5 THE FMS ENVIRONMENT

Automated or flexible manufacturing systems are rarely used in
an isolated environment. Rather, they are often part of a larger
production facility. This larger facility usually contains many
conventional production processes. Owing to the special nature
of FMS, there is a need for an interface between the FMS and its
neighbors. For example, FMS often operate on a batch-of-one

basis, unlike conventional systems. Finished items may come off the FMS at a rate that cannot be assimilated by the conventional facility. Conversely, facilities upstream from the FMS may need to produce raw workpieces to stock in order to maintain input to the FMS. Part of the interface of the FMS to the surrounding facility is incorporated in the FMS software, which can include algorithms for work release and for scheduling of workpieces from entry to completion. This software may reside on the FMS control computer, but could be complex enough to require a larger off-line computer for effective use. It should be noted that the algorithm used depends to a significant extent on the nature of the products made on the FMS. Some FMS are employed in a "make-to-stock" context where the timing of finished parts is not critical. Other FMS are designated to provide parts for immediately following assembly operations, in which case the movement of workpieces from work release to unload from the FMS must be done in a carefully controlled way.

Given some definitive idea of the FMS design (where "definitive" does not necessarily imply that the design is completed in detail), it is essential in the corporate world to determine the economic costs and benefits expected to be obtained from the system. Systems as complex as an FMS are not amenable to the conventional approaches to economic analysis. The essential incompatibility to conventional economic analysis is not just due to the system complexity. More often, current analyses use manpower rates as a basis for their calculations. FMS require less direct labor for their operation, but may require more skilled labor to support their operation. The peculiar aspects of FMS have given rise to new approaches to economic analysis. These approaches employ simulation and n-o-q models of the FMS as a part of the analysis technique. Several kinds of cost/benefit determination schemes are discussed in Chapter 13.

7.6 INTEGRATION OF MODELING METHODS

Though the modeling schemes discussed in this book are often complementary in their use, they do currently stand apart from each other. There is a considerable need to integrate these tools into one framework. One effort aimed in that direction was the Integrated Decision Support System (IDSS) developed by the Air Force ICAM program [7]. In IDSS, performance evaluation tools as well as economic modeling tools are loosely integrated under one umbrella language called DSL (presumably meaning Decision Support Language).

There is promise that the rapidly developing area of artificial intelligence (AI) software will provide a framework for a more integrated set of FMS design tools. Some recent developments toward incorporation of AI modeling tools into the FMS design context are discussed in Chapter 14.

REFERENCES

1. J. Solberg et al., *Factories of the Future: Defining the Target*, CIDMAC report, NSF grant MEA 8212074, Purdue University (1985).

2. A. D. Hall, *A Methodology for Systems Engineering*, Van Nostrand, Princeton, NJ (1962).

3. J. E. Gibson, "A philosophy for urban simulation," *IEEE Trans. on Systems. Man. and Cyber.*, 2:129 (1972).

4. A. P. Sage, *Methodology for Large-Scale Systems*, McGraw-Hill, New York (1977).

5. M. Barash, J. Talavage, et al., "Optimal Planning of Computerized Manufacturing Systems," Seventh NSF Grantees Conf. on Production Research and Technology, Ithaca, NY (1979).

6. *The Flexible Manufacturing Systems Handbook*, The Charles Stark Draper Laboratory, Cambridge, MA (1982).

7. *Integrated Design Support System (IDSS), Build 1*, AFWAL-TR-85-4017, Materials Laboratory, Wright-Pat. AFB, OH (1985).

8
Simulation Modeling

8.1 THREE APPROACHES TO SIMULATION MODELING FOR FMS

Computer simulation is a modeling tool that allows one to describe objects in the real world, such as workpieces, as words in a computer. By, in effect, moving these words from place to place in computer memory, the computer simulates workpieces moving from machine to machine in the real-world facility. Because the computer can perform the moves very rapidly even for large numbers of objects, the complex behavior of a system as complicated as an FMS over a time period of a day can be described in a matter of a few minutes. During the same time period, the simulation program can also be counting the number of workpieces processed by each simulated machine, the number waiting for each machine at each time instant, and so forth. These collected data are reported to the simulation user as the output of the simulation. Such data provide information on the degree to which the machines were utilized, the time spent waiting by the workpieces, as well as many other facets of the FMS performance.

For most, but not all, kinds of simulation, the prospective user must describe to the computer those objects and any relationships between them that describe the real-world-system behavior. This description is known as the simulation model. The performance data obtained from a simulation are a credible description of the system behavior provided that the detailed behavior of the FMS has been properly represented in terms of computer words and their transitions (i.e., a "good model" has been used).

Simulation, used for description of system performance, is an important modeling tool for several reasons [1]. Most important, it can provide insights regarding behavior of systems that are still in the design stage, thus saving the possibly high cost of discovering errors after construction. A similar benefit holds for study of existing systems which may not be tampered with, but which seem to need improvement (e.g., an inefficient manufacturing system that is currently producing valuable parts). A secondary benefit of simulation modeling is that it may force the user to think about the system in greater detail than was the case before the use of this tool. Often, this effort of carefully defining the system provides insights to system behavior even before a computer run is made. Another cited benefit of simulation modeling is that it can improve communication about a system among diverse groups of people working on its design (or redesign). This potential benefit is to a large extent dependent on the form of the model that the simulation procedure uses, as we discuss later.

From the user's standpoint, there are at least three approaches to developing a simulation model, as summarized in Fig. 8.1. A popular approach is that of "network" or graphical models, where some objects (such as machines) may be represented by graphical symbols placed in the same physical relationship to each other as the corresponding machines are in the real system. Newer simulation procedures accept these "layout" descriptions of factory facilities from a CRT screen as a major part of the model specification. At least the graphical aspects of this kind of model are relatively easy to specify, and once completed, they also provide a communication vehicle for the system design which can be readily understood by a variety of people. Network simulation procedures that have been widely used for models of manufacturing systems include GPSS [2], SIMAN [3], SLAM [4], GEMS [5], and CAPS/ECSL [6]. This type of procedure is also the most representative of those that have been implemented on microcomputers.

The network models were developed to ease the process of modeling, but they can still be restrictive in their capability to represent the real world since they provide only a small number

Approach	Characteristics
Network	Robust modeling capability Relatively easy to use Requires some programming ability Implemented on personal computer
Base	With sufficient effort, has most general model-ing capability Requires considerable programming ability Long training period for proficient use
Data-driven	Easy to use Requires no programming ability In conjunction with databases, can provide automated simulation Is not robust re modeling capability

Fig. 8.1 Three approaches to simulation modeling.

of graphical symbols to do so. The most general modeling capability is provided by a programming language (such as FORTRAN).
The earliest simulation programs consisted essentially of a set of
FORTRAN subroutines in terms of which the user constructed the
model (in, for example, the GASP language [7]). More recently,
other languages, such as SIMULA [8] and SIMSCRIPT [9], have
provided a framework for simulation modeling. Although these
frameworks for developing models are quite unrestrictive in their
representation power, they are also relatively difficult to use except for those versed in the base programming language.

A third kind of simulation procedure strives to accomplish simulation without programming. These simulation procedures may
be referred to as data-driven. The model for such a procedure
consists of mainly numerical data such as might be available from
a database in a manufacturing facility. For example, the data
could include the number of machines in the modeled facility, the
number of parts to be made there, the operation times for each
part on each machine, the route of the parts through the machines, and so forth. A list of this numerical information is presented to the computer as the model, and there is no need for the

user to understand a programming language or even to learn a set
of graphical symbols. In fact, ultimately, the user need not be
involved at all since the data could be automatically extracted
from a database. Though these data-driven procedures represent the ultimate in ease of use of simulation, they are more restricted than network models in the robustness of their modeling
capabilities. Currently available data-driven simulation procedures include GCMS [10], MAST [11], MAP/1 [12], and GFMS
[13].

All three kinds of simulation procedures require that the user
provide data for the modeling process. The type of data required
may be a little different in each case, but it is important to note
that the data often concerns rather detailed aspects of the modeled
system. This requirement to supply more detailed data for model
execution is one distinguishing feature between simulation and
network-of-queue (n-o-q) models (discussed in the next chapter).
Obtaining this detailed information may involve making (sometimes
difficult) choices by engineering staff during a design or may im-
pose a burden on a supervisor's time and data collection abilities
when modeling a system currently in operation. So the collection
of detailed data must be seen as one of the costs associated with
simulation modeling. On the other hand, regular use of simulation could spur the deployment of data collection and updating
procedures which could be of great benefit to the company in
ways perhaps unrelated to the simulation activity.

8.2 NETWORK SIMULATION MODELING

In this book, we are mainly interested in those real-world systems which can be described in terms of movement and processing of discrete objects. For example, in a manufacturing system,
we see workpieces or components moving about from one work
area to another. And in a factory communication system, we can
imagine messages being routed from one way station to another,
with some messages waiting while other (perhaps higher-pri-
ority) ones are being processed.

The network simulation languages have special constructs for
representing essentially five major aspects of movement and processing of objects [14]. That is, the objects are modeled as *arriving*, possibly *waiting* for processing, being *processed*, perhaps being *routed* to other processors, and finally *departing*
from the system that is the focus of the model. Since we are

interested in the behavior of the system as it processes the objects, there is a sixth aspect of simulation which is indispensable,
namely *data collection* on the system behavior. We can take these
six aspects to describe the types of real-world problems that network simulation may assist in solving.

Network languages were introduced by Gordon [15] in 1961
with his GPSS language. It was the most widely used simulation
language in the United States [16] and is still frequently used to
model manufacturing systems. In Europe, graphical models based
on Tocher's "wheel charts" [17] have been in use since 1969 [18].
The CAPS/ECSL language is based on the "activity cycle" dia-
grams that evolved from wheel charts.

Consider an example of the use of network languages as adapted
from Schriber [2]. The system of interest is that of an automated
work cell to produce machined parts. At this station, the part is
to be machined and subject to in-process inspection on the worktable. If the part is out of specification after the operation, it is
routed to a rework area. After rework, the part is returned to
the original station for inspection only. This situation is depicted
in Fig. 8.2, where circles represent parts.

To model this production system, we note that raw parts will
arrive to the work cell (from casting, say) with an interarrival
time between 4 and 8 minutes. Two machines are available to process parts at the station, each taking between 6 and 8 minutes for
operation and inspection. If both machines are busy, arriving

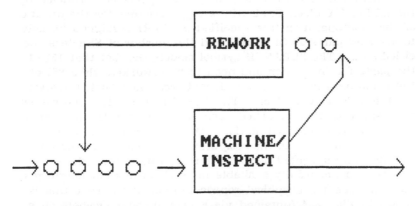

Fig. 8.2 Schematic of automated work cell model.

(or reworked) parts wait to be processed. About 85% of the parts complete the initial machining successfully and are routed to the packing department. The other 15% are routed to the rework area, which has one machine to do the rework. The rework process requires between 20 and 25 minutes.

Engineers designing the work cell suspect that the two machines will not be able to process the parts as fast as they arrive. The objective of the simulation analysis of the work cell is to determine how much floor space must be set aside to hold the parts waiting for processing and for rework. Our construction of a simulation model to represent this type of system consists of specifying "blocks" for each of the basic aspects of object movement. The blocks used in this example are taken from the MicroNET simulation language [19] and are shown in simplified form in Fig. 8.3. For each block, certain pertinent parameters, such as "duration," for an activity may also be specified, as seen in Fig. 8.3. The model that results from use of Micro-NET is in the form of a "network" of blocks that would appear as in Fig. 8.4.

Since the representation in Fig. 8.4 takes the form of a network of blocks, we refer to the model as a network model and to these languages as network languages. As indicated by its name, the CREATE block called CASTING_ARRIVAL represents arrivals of raw parts to the work cell. The ASSIGN node, OP_TIME_SPEC, specifies that newly arriving parts will be both machined and inspected and assigns the time for the combined activities to the part's information tag called ATRI. The SWAIT blocks represent the waiting areas for parts that cannot be processed owing to busy servers. The machining/inspection process is modeled by the ACTIVITY block CELL_OP, with parameters for the duration of the total operation time specified by ATRI (assigned initially to arriving parts as OP_TIME) and the number of SERVers specified as 2. The PBRANCH symbol models the fact that 15% of the parts fail inspection and need to be reworked while 85% of the parts pass inspection and depart from the modeled system via the TERMINATE block. The ASSIGN block after the rework node specifies that reworked parts only undergo the inspection activity at the work cell station. Once this network of blocks has been translated to the computer, the behavior of the work cell over any specified period of time can be simulated.

There are currently available network simulation languages that can accept a graphical representation of the type that is shown in Fig. 8.4 (obtained via a user placing symbols on a

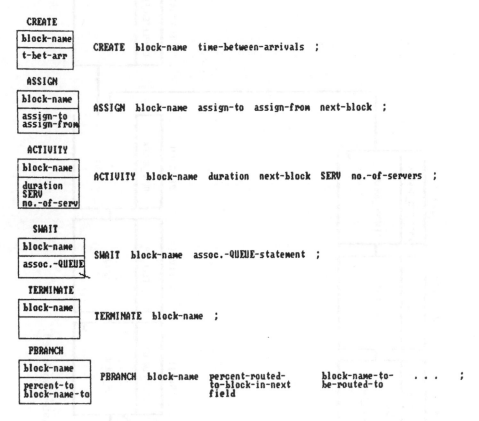

Fig. 8.3 MicroNET blocks and statements.

CRT screen) and, with additional information, provide a completed simulation of the modeled system (e.g., TESS [20]). For some other languages, each block must be described or translated into a statement or line of text according to a strictly specified format. A sample of those statements for our work cell model is shown in Table 8.1. Note that the blocks are described by statements listed under "(Network)." Each such statement begins with the block-type identifier. This is followed by the specific block's name, after which the block parameters are specified (each parameter is separated by spaces). In addition to the block statements, note that other statements listed under "(Definitions...)" are also required in order to specify, for example, the time OP-TIME (in

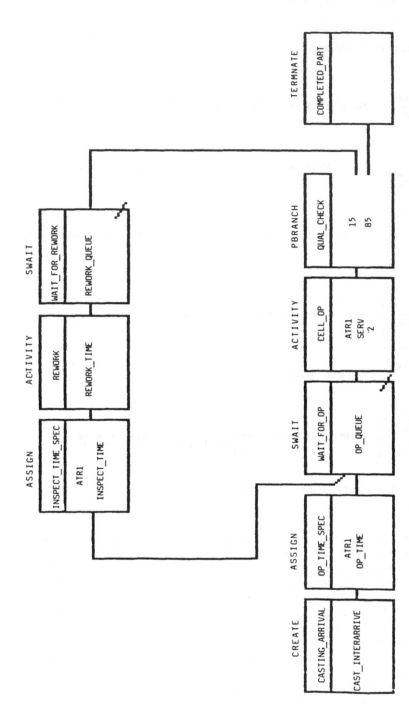

Fig. 8.4 Network model for automated work cell.

190

Table 8.1 MicroNET Statements for Automated Work Cell Model

(Definitions of uniform distributions and queue sizes in model)

UNFRM CAST_INTERARRIVE 4 8;

UNFRM OP_TIME 6 8;

UNFRM REWORK_TIME 20 25;

UNFRM INSPECT_TIME 2 3;

QUEUE OP_QUEUE 10;

QUEUE REWORK_QUEUE 10;

(Network)

CREATE CASTING_ARRIVAL CAST_INTERARRIVE;

ASSIGN OP_TIME_SPEC ATR1 OP_TIME;

SWAIT WAIT_FOR_OP OP_QUEUE;

ACTIVITY CELL_OP ATR1 , SERV 2;

PBRANCH QUAL_CHECK 15 WAIT_FOR_REWORK

                        85 COMPLETED_PART;

TERMINATE COMPLETED_PART;

(Subnetwork for rework: returns part to WAIT_FOR_OP)

SWAIT WAIT_FOR_REWORK REWORK_QUEUE;

ACTIVITY REWORK REWORK_TIME;

ASSIGN INSPECT_TIME_SPEC ATR1 INSPECT_TIME WAIT_FOR_OP;

---

this case, as a statistical sample from a uniform distribution between 6 and 8 minutes). Furthermore, the definitions given in the QUEUE statements specify the waiting areas as having 10 spaces for parts. The input processor of the simulation program reads this set of statements and creates a memory image of the model from it. Given this memory image, the user is ready to perform a simulation of system behavior over some time frame.

Because of their relative ease of use and relatively robust modeling capability, network languages have often been used to simulate the behavior of FMS. The languages that have been most often used in this regard are GPSS, SLAM, SIMAN, GEMS, and CAPS/ECSL. The method of use for the first four mentioned languages is quite similar for each, namely:

1. Sketch the model in terms of the blocks supported by that language.
2. Translate that sketch into a form the computer can accept (either lines of text or interconnected symbols on a CRT screen).
3. Execute the simulation and obtain the output.

The CAPS/ECSL mode of model description is different and, as has been mentioned, is based on an activity cycle diagram (ACD). This is a diagrammatic representation of a system in terms of activities and queues ONLY. Activities are shown as rectangles and queues as circles (names are given to each by the modeler). An ACD for the work cell being discussed is shown in Fig. 8.5. All system features that are to be modeled are termed entities (ECSL does not distinguish between entities, resources, and servers), and each class of entity is modeled as a closed-loop sequence of alternating activities and queues. For resource- and server-type entities (such as machines) that stay within a system, a closed loop is logical. For entities that pass

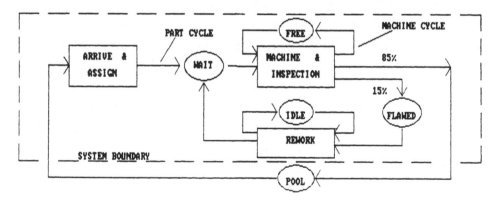

Fig. 8.5 Activity cycle diagram for work cell.

through a system (such as parts), the loop is closed by recirculating entities through a queue lying outside the model boundary (e.g., POOL in Fig. 8.5). Entities come together in activities and then separate into their own queues ready to take part in the next activity.

The level of detail shown in an ACD is much less than in most other symbolic representations of models used by other network languages. In deriving the diagram, the logical decisions must be considered, but are not explicitly represented. Once the ACD is available, the user can specify the model to a computer via CAPS, which has an interactive dialogue format. In this dialogue, the CAPS program inquires from the user as to which entities are to be represented, which activities and queues they pass through, the activity durations, the branching and other logic, and the output required. The program then proceeds to create an ECSL program of the situation to be modeled. This is presented to the user for approval. Refinements to the logic can be added by editing the program at this stage. CAPS does not have to be used as a front-end processor, as the simulation can be coded directly into ECSL.

Figure 8.6 shows the ECSL code for the machining and inspection activity. The ECSL language words are shown underlined; the others are chosen by the user to be as meaningful as possible. The CHAIN..OR words are equivalent to "either..or" or the "if..then..else" constructs of other languages. In this example activity, they enable the logic of deciding the type of part entering the activity, and branching on leaving, to be coded.

The "interview" approach to simulation modeling has in fact been extended to other languages (e.g., GASP, SIMSCRIPT) by Mathewson [21]. Currently, the interview approach to modeling is used to best advantage for initial modeling efforts. For complex systems, it suffers from an inability to obtain an expression of complicated decision-making procedures directly from the user. Thus, for an FMS that incorporates a complex material-handling logic, for example, the interview would be strained in order to obtain and represent that logic from the user. This difficulty is ameliorated with CAPS, as the ECSL code that is generated can be refined later by editing.

The above-perceived shortcoming of the interview approach to modeling, namely representing complex decision making, is to some extent shared by all the network languages. Though the blocks or constructs for representing movement of objects have

BEGIN MACHINE AND INSPECT ACTIVITY

FIND FIRST MACHINE A IN FREE

FIND FIRST PART B IN WAIT

CHAIN

       DURATION = INSPECTIME OF PART B

       TYPE OF PART B EQ FROMREWORK

       PART A FROM WAIT INTO POOL AFTER DURATION

       OR CHAIN

              DURATION = OPTIME OF PART B

              RANDOM(100,S) LE 84

              PART B FROM WAIT INTO POOL AFTER DURATION

              OR PART B FROM WAIT INTO FLAWED AFTER DURATION

MACHINE A FROM FREE INTO FREE AFTER DURATION

REPEAT

Fig. 8.6 ECSL code for the activity MACHINE & INSPECTION in Fig. 8.5.

been developed to an advanced state of refinement, the constructs that provide for logic in decision making are usually of the more primitive types (e.g., AND, OR, arithmetic relations). The modeler must then develop complicated decision-making schemes from these basic elements, as illustrated by the following example.

Consider the representation of a job shop situation in which many part types are to be produced, with each part type having a unique route through the shop. In most network languages, a part route is usually specified by a series of interconnection branches, with one branch between each two processing blocks

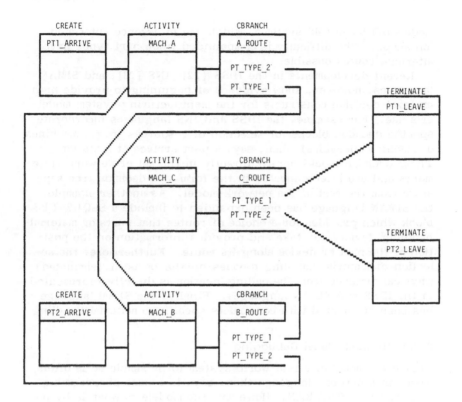

Fig. 8.7 Network model for three-machine shop.

on the route. If the modeled system is to process several part
types, then each processing block that works on more than one
part type will have more than one interconnection branch departing from it. One of the branches is taken conditionally
based on some property or attribute of the part. For example,
in the MicroNET simulation language, representation of part
routing for two part types through a three-machine shop might
appear as in Fig. 8.7. In that model, the CBRANCH block is
used to conditionally branch parts based on their type. Part
type 1 follows the route through machines A, then B, then C,
whereas part type 2 is planned to be machined at station B, then
A, then C. This "decentralized" means of representing routes
introduces considerable difficulty into modeling (and revising

models of) large FMS systems, which are in essence automated job shops. The difficulty is compounded when part types have alternate routes possible.

Recent developments in the IDSS [22], INS [23], and SIMAN languages, however, show evidence of beginning to provide higher-level-decision constructs for the manufacturing system modeler's use. For example, the IDSS and INS languages incorporate specific decision blocks for selection of resources (e.g., machines or machine operators) when, say, a part arrives at a station. These decision blocks appear directly in the set of network statements and are interconnected in the form of a decision tree separate from the rest of the network model. As another example, the SIMAN language has been expanded to include a SEQUENCES block which provides for a choice of routes that parts or material-handling devices can take and provides information on the position of the part or device along its route. Furthermore, the selection of material-handling devices on the basis of (simulated) physical distance from the calling location is directly represented in the IDSS and SIMAN languages. Finally, the SLAM language has been augmented with constructs specific to material handling.

8.2.1 Network Model Debugging

Creating a model of a real-world system to be simulated is often easier in a network language than, say, in a base programming language like FORTRAN. However, the modeler's work is by no means done when the model is sketched and translated to a form that the computer can accept. Ahead lie the important tasks of verifying that the model indeed accurately represents the real-world system and modifying the model if it fails to do so. This "debugging" process can be time-consuming and tedious, especially if attempted on a mainframe computer with long turnaround times. Recent developments in implementing simulation languages on the personal computer promise to alleviate many of the problems associated with the debugging process. Regarding the network languages, microcomputer (IBM-PC) implementations have already been marketed for the GPSS, SIMAN, SLAM, and CAPS/ECSL programs. Except for GPSS/PC [24], the PC versions are fully equivalent in modeling power to their mainframe counterparts. The GPSS/PC program is interesting in that it utilizes fully the fact that the PC is "owned" by the user and so continually available for user interaction. This interactive nature of GPSS/PC provides for a more convenient debugging process by the modeler. (It is anticipated that the other languages will also introduce interactive debugging capabilities.)

Let's consider the debug process in more detail. On a mainframe, the modeler who has translated the model to the computer then submits it for execution. Errors of verification may afterward be noted from the simulation output, in which case the model would need to be modified. For example, a typing error may have been made when specifying the processing time at a modeled machining station. This error would show in the simulation output data for that station. Usually, correcting an error entails that the modeler enter into a separate editing environment. When editing is complete, the model will generally need to be retranslated and then rerun. The behavior of the (possibly modified) model cannot be observed until the run is complete and the output returned. In the interactive PC environment, however, this debug-edit-run cycle can be accomplished in a more convenient way.

For example, editing of the model in GPSS/PC is done in the same integrated environment as model execution. Syntax error checking is automatic while doing the edit in GPSS/PC. (Syntax errors are those errors such as mentioned above which may result from mistyping the description of a block in its very strictly defined format.) After editing, the model behavior can be monitored as the run proceeds in several ways. By use of a STEP command, the modeler can move the model one (or more) steps at a time. The model status can be examined at the end of each of these steps. Another, more visual way to display the progress of the simulation is with the PLOT feature, where a pertinent variable of the model (e.g., waiting-area usage, machine utilization) is plotted on the CRT screen as the simulation proceeds. If this variable attains undesirable values, the simulation can be interrupted at any time and the reason for the behavior can be investigated. (An increasing number of simulation packages offer plots and more elaborate graphical output as the simulation proceeds. The use of graphics is considered again in Chapter 11.)

In addition to the advantages to the experienced modeler for debugging, this interactive environment could also be useful for education and training purposes. The ability to step through the modeled system behavior one step at a time allows showing the detailed system behavior in small assimilable increments, as well as displaying some of the basic operations of the simulation process itself.

In summary, the network languages provide a relatively easy-to-use, rather general representation or modeling capability. The established tradition of using symbols to represent modeled concepts makes these languages ideal for a CRT graphics interface to input model data. The wide availability of graphics capabilities on microcomputers along with the ease of use of network languages has thus led to their being a favorite for implementation in that environment. An added benefit of microcomputer implementation results from the interactive nature of single-user computers. The main need for further development of these languages lies in improving their ability to represent complex decision making.

8.3 DATA-DRIVEN SIMULATION PROCEDURES

The network simulation procedures require that the prospective user understand at least a basic core of symbols representing fundamental aspects of the real world. Moreover, that user must be able to translate the symbolic sketch of the real-world system to the computer in a nearly error-free way. Finally, the modeler needs to iterate in the debug-edit-run cycle until the model is verified. Only at this point can the simulation be used for design or redesign purposes in a credible fashion. Potential simulation users such as manufacturing supervisors generally do not have the time that it takes to accomplish this entire process. Even an industrial engineer member of the supervisor's staff might be able to devote only limited time to the modeling effort. These time pressures have motivated the development of a "programming-less" approach to simulation. This type of simulation is also referred to as data-driven.

The model for a data-driven simulation procedure consists only (or mainly) of numerical data. That information usually represents, for example, a simple count of machines in a system, or a table of operation times for each process on the route of a given part type. The nature of this information is such that, if it were collected in the factory information system, it would only be necessary to access it and place it in proper format in order to run a simulation of the corresponding real-world system. This concept is quite close to *automated* simulation. Some pertinent aspects of automated simulation are discussed in Chapter 11.

Among the earliest data-driven simulators were the GENMOD program developed by the Army [25] and the GALS assembly

line simulator developed at IIT [26]. The first such program for
FMS was developed at Purdue; it is called the General Computerized Manufacturing System (GCMS) simulator [10]. The GCMS
program will be our focus for examples and detailed discussion
later. Several packages, including MAST and GFMS, are derivatives of the GCMS program. More recently, a commercial simulator named MAP/1 has been marketed which employs the data-
driven approach in its user orientation.

Because of their "ultimate" ease of use, the data-driven simulation procedures have a special character. In a sense, the detailed model that is the focus of the simulator is "built in" to the
data-driven simulation procedure. As an illustration of this,
consider the GCMS simulator. The eight key FORTRAN subroutines in GCMS represent changes that take place in the FMS during the following:

1. The operation on a part at a station
2. The completion of the part's operation
3. Movement of the part into a parts-completed queue if one exists
4. Call for a material-handling (mh) device if such devices are
   in the system; otherwise place the part on a conveyor
5. Movement of a part onto an mh device other than a conveyor
6. Movement of the mh device carrying the part to the part's
   next station
7. Movement of the part into the next station's waiting area if
   one exists
8. Movement of the part from the waiting area into the processing station when the station is available and the part meets
   the selection criteria

These subroutines are entirely generic and do not distinguish,
for example, whether the station is a machining station or a metal-
forming station or even an inspection station.

The only assumption made by GCMS is that all the actions of
the system that are important to its performance are described
by a sequence or subsequence of the eight basic subroutines.
Then the input data to GCMS is greatly simplified since it only
acts to parametrize the existing subroutines. For example, one
item of the input data indicates the number of places for part
storage in the parts-completed queue at a given station. If that
number is zero, the generic subroutine associated with moving a
part into the parts-completed queue will be bypassed for that

station. Another item of input data to GCMS is the mh device
speed. In essence, the "equations" of motion of these devices
are built into GCMS, and so the program needs just the (average) speed and the distance between stations (also provided by
input data) to calculate travel time. By comparison with network
simulation procedures, it would not be amiss to view data-driven
simulation procedures as having eight "blocks" corresponding to
the eight subroutines mentioned above. The numerical data input to the data-driven simulator essentially creates (automatically)
a kind of "network" model which is used to describe performance.

As an example of the use of a data-driven simulator, imagine
that you are called on to evaluate the preliminary design of an
FMS. For ease of illustration, this FMS is to produce only one
product type. This product, called a widget, requires four operations as well as fixturing and defixturing operations. Both
types of fixturing occur at one load/unload station. The system
has already been designed to have a drilling machine, a lathe, a
vertical turret lathe, and an inspection station, all laid out as in
Fig. 8.8.

The first operation on the widget is to be performed on the
drilling machine, followed by an operation on the lathe, then another lathe operation on the VTL, and finally inspection of the
part is performed. The VTL is set up, in fact, to perform both
lathe operations if necessary. The operation times are as follows:

1. Fixturing   10 minutes
2. Drilling   20 minutes
3. Lathe   10 minutes — (if performed on VTL, time
                         is 15 minutes)
4. Lathe (VTL)   20 minutes
5. Inspection   5 minutes
6. Defixturing   5 minutes

The parts-waiting and parts-completed areas at all the stations
except the VTL have space for two parts. At the VTL station,
those areas can hold only one part.

The mh system is assumed to be composed of two tow carts.
Their average speed is 2 ft/minute, and they can travel in either
direction along the track.

In order to limit congestion in the system, only four pallets
are available for fixturing parts.

The GCMS input data (or model) for this example consists of
rows of numbers, each row representing the numerical data either

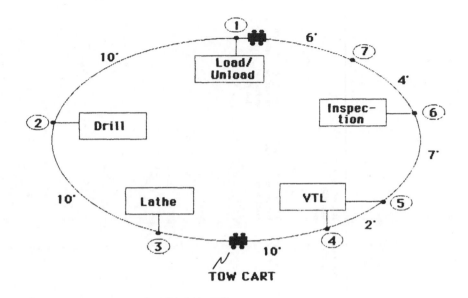

Fig. 8.8 Layout of widget example.

for the part types to be produced (including process plan and operation times), for the fixture or pallet types available for the parts, for the work stations in the FMS, or for the mh system characteristics. The format for each type of data row is given in Table 8.2. Since each row consists of numbers separated by commas (except for the first field in each row which contains an identifier), this is called a free-format data input. (A fixed-format data input would require each number to be right-justified to a particular column, as in some FORTRAN input formats.)

The GCMS input data for our example FMS is shown in Table 8.3. The first line describes some system characteristics, including the name of the product (WIDGET), the number of part

Table 8.2 GCMS Free-Form Input Format

1	2	3	4	5	6	7	8	9	10	11
SYSTEM NAME	, Total No. of operations for all part types	, No. of part types in system	, No. of pallet types in system	, Max. no. of workpieces in system	, No. of workstations in system	, No. of cart types in system	, No. of track decision points			
OPERATION NAME	, Operation number (consecutively from 1 for all operation numbers)	, Decision rule for finding an alternative station 1-Highest priority station. The station that appears first on the input alternative list is considered to be of the highest priority. 2-Idle station first. 3-User's function	, First alternative work station no.	, Operation time	, Second alternative work station no.	, Operation time	, Third alternative work station no.	, Operation time	, Fourth alternative work station no.	, Operation time

For each operation

WORK STATION NAME

For each workstation:

1	WORK STATION NAME
2	Station number (consecutively from 1 for all station numbers)
3	No. of operations performed at this station
4	No. of on-shuttle queue positions
5	No. of off-shuttle queue positions
6	No. of generalized queue positions
7	Decision rule for picking up a part from queue 1-FIFO 2-Highest priority index 3-User's function
8	Queue to track move time
9	Queue to machine move time
10	Track decision point no. for on-shuttle
11	Track decision point no. for off-shuttle
12	Breakdown distribution code
13	Breakdown distribution mean
14	Breakdown distribution standard deviation (if applicable)
15	Breakdown distribution minimum value (if applicable)
16	Breakdown distribution maximum value (if applicable)
17	Repair distribution code
18	Repair distribution mean
19	Repair distribution standard deviation (if applicable)
20	Repair distribution minimum value (if applicable)
21	Repair distribution maximum value (if applicable)
22	Maximum time allowed before part moves off failed station
23	Alternate station # for this station's operations

DISTRIBUTION CODES

1. Exponential
2. Normal
3. Uniform
4. Triangular
5. Lognormal
6. Erlang
7. Poisson
8. Gamma
9. Beta

(continued)

Table 8.2 (Continued)

PART TYPE NAME	No. of part type (consecutively from 1 for all part types)	User-assigned number for part type (may be same as field 2)	Max. no. of workpieces of this type in the system at one time	Decision rule for finding next operation 1-Pick an operation according to routing order. Number should not be repeated in each part operation route 2-User's function	Decision rule for finding a cart 1-Find an idle cart with the lowest cart no. 2-Find a cart with nearest destination. 3-Conveyor system 4-User's function	Priority	Production Target	No. of operations required for this part type	Operation number sequence (max. = 30)
	1	2	3	4	5	6	7	8	9
For each part type	1	2	3	4	5	6	7	8	9	10...39

PALLET TYPE NAME	Pallet type number (consecutively from 1)	No. of this type's pallets available	Station no. at which all pallets of this type are initially located	Part type no. which it can be used for (max. = 10)
	1	2	3	4
	1	2	3	4	5...14

	1	2	3	4	5	6	7	8	9	10	11
For each mh device type	mh DEVICE TYPE NAME	mh device type number (consecutively from 1 for all device types)	No. of this type available	No. of pallet positions on each device of this type	Speed over decision points	Decision rule for scheduling movement. 1-Schedule movement according to activity assignment 2-User's function	Feasible track decision points (max. = 46) 7...52				
For each decision point	DECISION POINT NAME	Decision point no. (consecutively from 1)	Workstation no. (if there is none, specify zero)	Conveyor speed (for cart system (specify zero)	First successor decision point number	Distance to the successor	Second successor decision point number	Distance to the successor	Third successor decision point number	Distance to the successor	Fourth successor decision point number , Distance to the successor
	1	2	3	4	5	6	7	8	9	10	11
For each mh device	mh DEVICE NAME	Initial decision point no. for device 1	mh device type for device 1								
	1	2	3								
For each SIMULATION RUN	SIMULATION RUN NAME	Simulation end time	Simulation trace start time	Simulation trace and time	Random number seed identifier (1...6)						
	1	2	3	4	5						

Table 8.3 GCMS Data Input for WIDGET Example

WIDGET, 6, 1, 1, 100, 5, 1, 7*

PALLET LOAD OP, 1, 1, 1, 10*

DRILL OP, 2, 1, 2, 20*

LATHE OP, 3, 1, 3, 10, 4, 15*

LATHE OP (ON VTL), 4, 1, 4, 20*

INSPECTION OP, 5, 1, 5, 5*

PALLET UNLOAD OP, 6, 1, 1, 5*

PART TYPE 1, 1, 1, 100, 1, 1, 1, 10, 6, 1, 2, 3, 4, 5, 6*

PALLET TYPE 1, 1, 2, 1, 1*

LOAD/UNLOAD STATION, 1, 2, 0, 0, 2, 1, 0.3, 0.3, 1, 1*

DRILL STATION, 2, 1, 0, 0, 2, 1, 0.3, 0.3, 2, 2*

LATHE STATION, 3, 1, 0, 0, 2, 1, 0.3, 0.3, 3, 3*

VTL STATION, 4, 2, 1, 1, 0, 1, 0.2, 0.2, 4, 5*

INSPECT STATION, 5, 1, 0, 0, 2, 1, 0.3, 0.3, 6, 6*

CART TYPE 1, 1, 2, 2, 2, 1, 1, 2, 3, 4, 5, 6, 7*

DEC PT 1, 1, 1, 0, 2, 10, 7, 6*

DEC PT 2, 2, 2, 0, 3, 10, 1, 10*

DEC PT 3, 3, 3, 0, 4, 10, 2, 10*

DEC PT 4, 4, 4, 0, 5, 2, 3, 10*

DEC PT 5, 5, 4, 0, 6, 7, 4, 2*

DEC PT 6, 6, 5, 0, 7, 4, 5, 7*

DEC PT 7, 7, 0, 0, 1, 6, 6, 4*

CART 1, 1, 1*

CART 2, 2, 1*

ONE 8-HOUR-SHIFT RUN, 480, 0, 50*

---

types to be produced (in this case, 1), the total number of operations to be performed by the FMS (6), and the number of workstations (5). The next six lines describe the operations on

the product. Note that in line 4 (labeled LATHE OP), that operation can be done on machine 3 (LATHE STATION, the third
station described below it) with a processing time of 10 minutes,
or at machine 4 (VTL STATION) with a time of 15 minutes. The
eighth and ninth rows describe the part and fixture data, while
the tenth through the fourteenth rows describe the machines or
workstations in the FMS. The remaining rows describe the mh
system, including the "decision points" where an mh device may
stop to either load, unload, or switch tracks (there are no track
switch points in this example). Note that decision points 4 and 5
are associated with one station (the VTL STATION). This means
that there is an area at this station for parts waiting to be processed that is separate from the area where completed parts wait
for their transport to the next operation.

GCMS needs only the numerical data in Table 8.3 to simulate
the FMS behavior and provide output for assessing its performance (in this case, for ONE 8-HOUR SHIFT). The performance
data one obtains from GCMS for this system are shown in Table
8.4. These data include information on part production, station
utilization (including the usage of waiting areas), and mh system utilization. Note, for instance, that no machine was utilized
more than 36%, and in the summary data for "operations performed," the operation that could be performed on either machine 3 or 4 (i.e., the LATHE or the VTL) was in fact performed all eight times on the LATHE. Some experimentation to
improve the performance of FMS using simulation will be described
in Chapter 11.

The price that is paid for the ease of use of data-driven simulation procedures is a considerable loss of generality of modeling power. Since so many modeling assumptions are built into
the simulator, these simulation procedures apply accurately only
to the systems that satisfy those (usually many) assumptions.
As an example, consider the GCMS simulator. Once a part enters a station for an operation, the program assumes all is ready
for that operation. In particular, the assumption is made that
the tools for the operation are accessible and in good repair. In
fact, for some proposed FMS, the tools (as well as the parts) are
to be delivered to the station. In these systems, the possibility
exists that the tool will not yet have arrived at the station when
the part is there. The GCMS simulator would have to be modified in order to model this situation. Past discussions with users
of GCMS indicate that each user has at least one or two aspects
of a proposed FMS in mind that would not be modeled directly
(i.e., without "tricks"), or without modification to the basic

Table 8.4 GCMS Output for Widget Example

Production summary for completed widgets

Part type	Parts sched	Parts compl	%	Ave	Min	Max
	Production			For parts completed time in system		
1	10	8	80.0	114.21	109.10	121.80
Total	10	8				

Operations performed

Operation number	Performed at station	Times started	Operation time
1	1	10	10.00
2	2	9	20.00
3	3	8	10.00
3	4	0	15.00
4	4	8	20.00
5	5	8	5.00
6	1	8	5.00
Total		51	

Station performance summary

Station number	Time busy	%	Time idle	%	Time down	%	% of time busy during time available
1	136.20	28.4	343.80	71.6	0.	0.	28.38
2	168.60	35.1	311.40	64.9	0.	0.	35.13
3	80.00	16.7	400.00	83.3	0.	0.	16.67
4	160.00	33.3	320.00	66.7	0.	0.	33.33
5	40.00	8.3	440.00	91.7	0.	0.	8.33

Shuttle performance summary

Station number	On-shuttle				Off-shuttle				Generalized			
	Queue size	Ave que	%	Max que	Queue size	Ave que	%	Max que	Queue size	Ave que	%	Max que
1	0	0.		0	0	0.		0	2	0.070		1
2	0	0.		0	0	0.		0	2	0.045		2
3	0	0.		0	0	0.		0	2	0.059		1
4	1	0.007		1	1	0.083		1	0	0.		0
5	0	0.		0	0	0.		0	2	0.094		1

Cart performance summary

Cart number	Cart activities								Total distance moved	No. decision points passed	No. of assignments	Initial assignments
	Move time	%	Waiting time	%	Down-time	%	Idle time	%				
1	188.00	39.2	17.80	3.7	0.	0.	274.2	57.1	376.00	0	25	25
2	126.00	26.3	10.00	2.1	0.	0.	344.0	71.7	252.00	0	16	16

program. As a result, a variety of GCMS programs designed for different purposes have evolved.

Modification of the GCMS program is rather difficult partly because it is written in unstructured FORTRAN. Yet, as indicated above, modifications are often necessary. To circumvent the modification problem, an experimental data-driven simulator has been developed at Purdue that is, in fact, based on a network simulation language [27]. This simulator for FMS (called PATHSIM) has been created in a modular fashion similar to GCMS (recall the eight basic actions). In PATHSIM, each of the basic actions is represented by a SLAM network rather than by FORTRAN code. These module networks can be perceived in graphical terms and thus still offer the ease of modeling and ease of modification of the network simulation procedures. Yet, the input to the simulator is in very much the same numerical format as for GCMS models. With this enhanced modifiability, it was possible to develop in PATHSIM a capability to model not only the FMS that GCMS could represent, but also those for which tool delivery is implemented, and it was possible to do so in a relatively short period of time.

In summary, the data-driven simulation procedures provide great ease of modeling, i.e., numerical data only (provided, of course, that such data are available). They therefore offer the best potential for automated simulation. These advantages are offset to some extent by their relatively nonrobust representation capabilities.

8.4 DATABASES FOR SIMULATION

Simulation may be seen as an input-output process where both the input and the output are constituted by sets of information. The input set of information represents the model, as well as control of the simulation run. The output set of information contains the data collected from the run. For FMS systems modeled via data-driven simulators, the input information usually consists of numerical data on parts, operations, machines, and mh subsystems. The potential to extract such numerical data from company databases for purposes of automated simulation seems strong, but such a practice has not often been reported in the literature.

The output information from a simulation run can be much more extensive than simply the collection of means and standard deviations of various queue and resource measures. For example, it

is conceivable to collect the entire state of the simulation model
at every time instant that the state changes. This set of data
would be completely comprehensive, though perhaps intractably
large. Some data collection schemes have been implemented where-
by it is possible to collect "snapshots" of the model state at strategic times and use these snapshots to reconstruct the remaining
information. The resulting ability to peruse simulation output at
the modeler's leisure for relationships that may not have been
foreseen when the model was run is the basis for packages such
as SDL [28] (SDL is now incorporated into the TESS program).
SDL can be used to collect simulation data into a relational database that may be subject to a wide variety of queries at any time
after the run. These queries, of course, include those relating
to the usual statistical measures. In addition, however, a kind
of interactive after-the-run analysis procedure can also be ef-
fected on the data in order to localize problem symptoms in, for
example, complex FMS control situations.

8.5 SIMULATION USING A BASE
PROGRAMMING LANGUAGE

The modeling restrictions imposed by the network and the data-
driven simulation procedures generally do not apply to simulation models written in a base programming language. Included
in this set of languages are FORTRAN, PASCAL, and C. However, other languages which could also be categorized as base
languages, but which contain procedures for simulation, include
SIMSCRIPT and SIMULA. The modeling capabilities of these
languages are limited only by the ingenuity and perseverance of
the modeler. A model constructed in one of those languages is
subject to nearly all the comparative criteria usually associated
with any computer program. For example, documentation of the
model's meaning (as opposed to syntax) requires the same considerations as documenting any other program in that language.
Modifications to a model's intent also engender the same advantages and difficulties as conferred by that language on nonsim-
ulation programs. The situation for SIMSCRIPT and SIMULA is
different, however, because these otherwise general-purpose
languages also provide special constructs peculiar to simulation
modeling. Simulation models constructed in these languages are
certainly more self-documenting than those developed in FOR-
TRAN, for example. Modification of simulation models in

SIMSCRIPT and SIMULA is more straightforward owing to the use of model-specific constructs.

Despite the extensive modeling capability of these two languages, they are not widely used for manufacturing system modeling. The reasons for this probably have little to do with the languages' capabilities, but rather have to do with the smaller number of people who know these languages well enough to use them. The training period for their expert use is considerably longer than for network or data-driven simulation procedures. More to the point, these languages are not taught frequently in engineering courses where prospective manufacturing system modelers would be exposed to them.

The situation with respect to FORTRAN is somewhat more ambiguous. Although not many models are constructed directly in FORTRAN, a considerable number of models are indirectly so constructed. This is because several of the network and data-driven simulation procedures are in fact written in FORTRAN. These include the SLAM, GCMS, MAST, GEMS, and MAP/1 programs. Since FORTRAN is the base language for SLAM and SIMAN, these languages moreover allow the modeler to combine network modeling with other aspects of the modeled situation written in FORTRAN code. Usually, the parts of the model written in FORTRAN represent aspects of the real-world situation that are not easily handled by the network constructs, such as complex decision making.

8.6 SIMULATION PROCEDURE COMPARISON FOR FMS DESIGN

The network simulation procedures and the data-driven ones presented above need not be seen as competing tools for design. If one considers the inherent properties of each, it becomes apparent that they are in fact complementary. The network languages are robust in their modeling capability and provide relative ease of use for modeling. In the early stages of design, these are quite advantageous characteristics since the system is not yet specified in detail and is subject to considerable change. As the design nears completion, however, or when the system is constructed and in operation, the desired improvements in performance may depend on some very detailed aspects of the system configuration. For the latter cases, a data-driven simulation procedure can be more appropriate since the system elements

are not being changed in a radical manner (presumably the configuration and control policies of the FMS have been stabilized),
and so the need for remodeling is significantly reduced. For example, for efficient operation of an FMS already in existence, it
may be necessary to "tune" the system with regard to the right
number of pallets to minimize congestion in the system. But this
number is usually just one of the data elements forming part of
the data-driven simulation procedure input.

In operation, the data-driven procedure offers the promise of
automated use of simulation. That is, such a simulation is used
not by a human, but by another computer program. For example,
the need for efficient and timely production implies that some control be exerted over the introduction of parts to an FMS. The
parts entering the system may be scheduled by a program that
includes a data-driven simulator. This "internal" simulator may
be used for several purposes, one of which is to estimate the performance of a candidate schedule generated by the scheduler program. Parameters of the real system that change from day to day
could be obtained for the simulation from an FMS database. Thus,
the system scheduler program would be employing a constantly up-
dated model of the FMS, where the updating function is automatic
and depends only on the factory database being properly maintained. A software product, called FACTOR, that operates along
these lines has recently been introduced [29].

The network and data-driven simulation procedures can be
viewed from another aspect, namely that of the organizational objectives and of the personnel in the FMS design group itself. If
such a group is to be called on to design FMS, say, for a large
corporation according to a "standard" configuration, then the
use of a data-driven simulation procedure could be most appropriate. On the other hand, if such a group can expect to be designing a quite different system each time, then a more flexible
tool such as network simulation would be advantageous, especially if that network simulator had the ability to interface with
a base programming language. (The latter feature could be used
to model FMS aspects that are unique to given manufacturing situations.) An even different perspective on the design group may
be taken if one considers that such a group may for some reason
have a high personnel turnover rate. In this case, continuity of
information flow over time is a problem for maintenance and reuse
of models, as well as for creating new models. Entering profes-
sionals may be familiar with simulation languages that are not yet
"accepted" by company management. One solution is the use of

data-driven simulators, for which the need to pass on information, the need for maintenance, and the time to learn the simulation package are minimal.

8.7 SUMMARY

Simulation has for a long time been used in industrial and manufacturing applications. Prospective users have influenced the development of such languages, particularly in the direction of making them easier to use for modeling purposes. This is manifest in the progression from base language modeling (e.g., FORTRAN) to network representations, culminating in data-driven simulation languages.

The progress in modeling capability has been accompanied more recently by a downsizing of the required computer to implement simulation. Many simulation languages are now available on microcomputers, especially for the IBM-PC. This single-user environment offers a degree of user interaction not previously available on mainframe computers.

Software associated with simulation languages, such as databases and statistical packages, has been the object of some development. The prospect of linking databases to simulation (especially data-driven simulation procedures) provides the potential for automated simulation. Statistical methods are usually associated with simulation to control the simulation runs (e.g., run length) and to analyze the output. The latter topic is considered further in Chapter 11, where we discuss the use of simulation procedures for detailed design and analysis of FMS.

REFERENCES

1. R. Shannon, *System Simulation: The Art and the Science*, Prentice-Hall (1975).

2. T. Schriber, *Simulation Using GPSS*, John Wiley, New York (1974).

3. C. Pegden, *Introduction to SIMAN*, Systems Modeling Corp. (1982).

4. A. Pritsker, *Introduction to Simulation and SLAM*, Halsted Press (1984).

5. D. Phillips and R. Heisterberg, *Definition, Development, and Implementation of a Generalized Manufacturing Simulator*, GEMS-3-77 report, Texas A&M University, December (1977).

6. A. Clementson, *The New Extended Control and Simulation Language*, Dept. of Engineering Production, Univ. of Birmingham, England (1973).

7. A. Pritsker, *The GASP IV Simulation Language*, John Wiley (1974).

8. O. Dahl and K. Nygaard, "SIMULA - An ALGOL-based Simulation Language," *Comm. of the Assoc. for Comp. Mach.*, *9*:9 (1966).

9. P. Kiviat, R. Villanueva, and H. Markowitz, *The SIM-SCRIPT II Programming Language*, Prentice-Hall (1969).

10. J. Talavage and J. Lenz, *General Computerized Manufacturing Systems (GCMS) Simulator*, NSF Report No. 7, NSF Grant No. APR 74-15256, August (1977).

11. J. Lenz, *The MAST User Manual*, CMS Research Inc., Oshkosh, Wisconsin.

12. L. Rolston, "Modeling Flexible Manufacturing Systems with MAP/1," *Proc. of Flexible Manufacturing Systems Conf.*, Ann Arbor, Michigan, August (1984).

13. The Charles Stark Draper Laboratory, *Flexible Manufacturing System Handbook*, Vol. 1, Nat. Tech. Info. Serv., US Dept. of Commerce, February (1983).

14. J. Talavage, "The PC Simulation Workstation," internal report, Purdue University (1985).

15. G. Gordon, *System Simulation*, Prentice-Hall (1978).

16. H. Kleine, "A Second Survey of User's Views of Discrete Simulation Languages," *Simulation*, *17*: 2, August (1971).

17. K. Tocher, "Some Techniques of Model Building," *Proc. IBM Scientific Computing Symp. on Simulation Models and Gaming*, IBM Corp., New York (1964).

18. P. Hills and T. Poole, "A Method for Simplifying the Production of Computer Simulation Models," *TIMS Tenth American Meeting*, Atlanta, October (1969).

19. J. Talavage and W. Lilegdon, *MicroNET, User's Manual for a Microcomputer Simulation Language*, Pritsker and Assoc., May (1983).

20. C. Standridge, "Performing Simulation Projects with the Extended Simulation System (TESS)," *Simulation*, December (1985).

21. S. Mathewson, "Simulation program generators," *Simulation*, December (1974).

22. *IDSS 2.0 Users Manual, IDEF2 Modeling*, IDSS Build 1 Final Report, AFWAL/MLTC, ICAM Program Library, Wright-Pat. AFB, OH (1984).

23. S. Roberts, *Simulation Modeling and Analysis with INSIGHT*, Regenstreif Institute, Indianapolis, IN, August (1983).

24. S. Cox, "The User Interface of GPSS/PC," Proceedings of Winter Simulation Conference, Dallas (1984).

25. E. Loniewski, *GENMOD - User Manual for Generalized Production Line Modeling Routine*, Rept. No. ARPAD-SP-78001, US Army, August (1978).

26. IIT Research Institute, *GALS - Generalized Assembly Line Simulator*, Chicago (1975).

27. J. Talavage and R. Mills, "PATHSIM, A Modular Simulator for the Study of Tool Movement in FMS," *Journal of Manufacturing Systems* (1985).

28. C. Standridge and A. Pritsker, "An Introduction to the Simulation Data Language," Proc. of Winter Simulation Conf., San Diego, pp. 617–620 (1982).

29. H. Grant, *FACTOR*, Factrol, Inc., West Lafayette, IN (1986).

9
Network-of-Queue Modeling

9.1 INTRODUCTION

Examples in the previous chapter (and in Chapter 11) indicate
that using computer simulation as a design or operational decision support tool for FMS implies that the characteristics of the
FMS are known in considerable detail. However, for an organization just contemplating the construction of an FMS, few details of the system will be clear. In fact, early on, even the
parts that will be made on the system may not be determined
with certainty. Thus, the FMS configuration could take on several equally satisfactory forms. A key issue, then, is how to
develop such incompletely specified systems into a more advanced
stage of design. We have already discussed the system engineering design process in which design is achieved by using iterative
analysis. A credible analysis tool for use in these early design
situations could save appreciable time, effort, and money in the
long run since it could focus design effort on "good" alternatives.
Such a tool must provide performance measures not unlike those
obtainable from simulation, but with the use of considerably less
detailed information. A class of models with the desired properties is available, having been developed from work on queuing

systems dating back to the early 1970s. These models deal with interconnection of queue stations and will be called network-of-queue (n-o-q) models.

As already noted for network simulation models, manufacturing systems can be fruitfully viewed from the perspective of material *movement*. With such a point of view, the major concern is with measurable criteria such as production rate, in-process inventory, and resource utilization. Emphasis on movement has already been shown to place importance on the concepts of service and queue (i.e., these are basic concepts of network simulation). Furthermore, a system object such as a part may engage in several consecutive service activities before leaving the system, and so there is an important notion of "routing" of these objects among the activities and their associated queues. These three concepts of service, queue, and routing or branching form the basis for n-o-q models.

9.2 RELATIONSHIP OF n-o-q MODELS TO SIMULATION

Before proceeding to the theoretical basis for n-o-q models, let us relate them back to network simulation models. From the basic MicroNET nodes described in the previous chapter, we can construct a model for an FMS that will parallel an n-o-q model for the same FMS. Consider a proposal for an FMS that is tentatively specified to produce two part types according to their already specified process plans. Imagine that a heuristic program (discussed in more detail in Chapter 10) for choosing machines, given the part information, has been used to select machines. At this point, then, the FMS is specified only in terms of the parts to be produced (and their rough process plan, which specifies just total time at each machine) and the number and types of machines that the parts are to visit. The final factor to be modeled is the movement of parts from station to station. Suppose we model all movement by a single activity (called the material-handling system, or mhs) with an activity time that is expected to be about the average move time. Of course, every part will undertake this activity as it moves from station to station. Our network simulation model would at this point look as shown in Fig. 9.1.

The model in Fig. 9.1 is not complete since the branching of parts from the mhs to the stations is not specified. In network simulation, this can be done in two ways, either by conditional or by probabalistic branching. If done via conditional branching, the type of part and its previous station could be used to

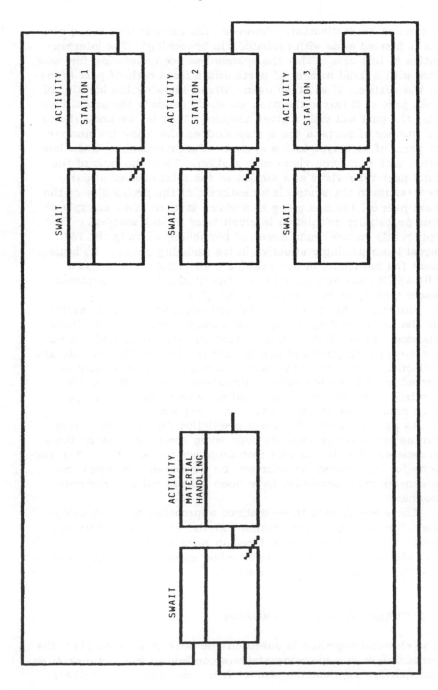

Fig. 9.1 Incomplete simulation model for closed queueing network.

determine its destination. However, the parallel with n-o-q models is instead made with probabilistic branching! The interpretation in this case is that the system has been operating for some
time with a fixed number of parts and a fixed ratio of part types
in the system. If so, then even *without* knowing the identity of
each part as it leaves the mhs, we could estimate the probability
that the part will visit a given station. That is, we know the total number of parts in the system and we also know the number
of parts of each type in the system. Furthermore, we know how
often each part type visits each station. Thus the ratio of the
total part-type-visits at a station to the total number of part-
type-visits in the system is an estimate of the probability of the
next part off the mhs going to a given station. (An example of
this probability calculation is given later in this chapter.) This
apparently rather crude means of branching parts in the FMS
model is surprisingly effective in its modeling power. To implement the branching in our example FMS model, we can use a
PBRANCH node between mhs and the stations. The completed
model then appears as shown in Fig. 9.2.

Simulating this network model will not give the same results
as the corresponding n-o-q model except under certain circum-
stances. In particular, imagine that the simulation model is run
for a very long period of time before the first statistical data are
collected. This places the model state into what is usually referred to as a "steady state" (discussed in more detail below).
Statistics collected from this point on would, on the average,
give results comparable to those of n-o-q models.

As we show later, the n-o-q prediction for production rate
can be surprisingly accurate even when some basic assumptions
associated with the analysis technique are not satisfied. The reasons for this robust behavior can be discussed only when the
n-o-q methods themselves have been considered in a rigorous
manner.

There are at least three distinct approaches to n-o-q analysis. Two have been used rather extensively to study FMS systems, and they will be considered in more detail. The three methods will be referred to as the classical approach, the mva approach,
and the operational analysis method.

9.3 CLASSICAL n-o-q APPROACH

The classical approach is based on the work of Jackson [1]. His
concern was with queueing stations configured in a network (i.e.,
the network representing travel from one station to another).

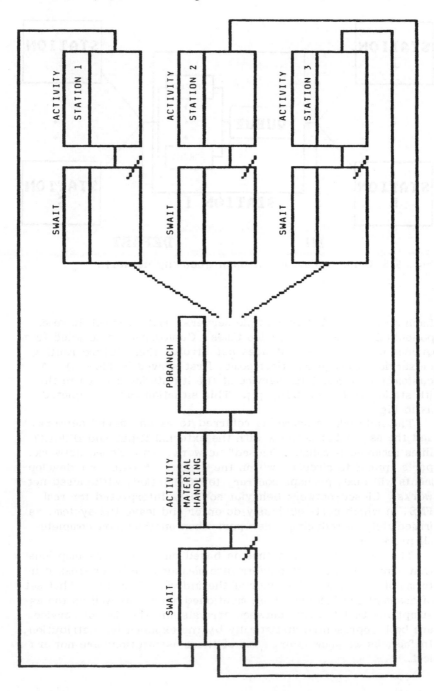

Fig. 9.2 Completed simulation model for closed queueing network.

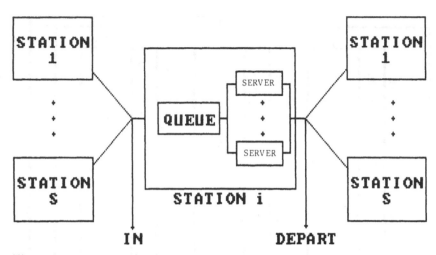

Fig. 9.3 General diagram for open queueing network.

Each station could have multiple servers, each of which have ex-
ponentially distributed service times. Customers are selected from
queues on any basis that does *not* involve their future routing
or service times (e.g., first-come, first-served is allowed). A
customer that completes service at the ith station may go to the
jth station with probability p_{ij}. This situation can be depicted
as in Fig. 9.3.

The network as shown is referred to as an "open" network,
and the associated network with the external INput and DEPART
lines removed is called a "closed" network. In a closed network,
parts appear to *circulate* within the system. Most of our developments will deal, perhaps contrary to expectation, with closed networks. Closed-network behavior could be interpreted for real
FMS, in which parts obviously do enter and leave the system, as
immediately introducing a new part when another part completes
all processing.

The classical n-o-q method is based on a set of assumptions
that allows the *network* performance measures to be expressed in
terms of the relative behavior of the *individual* queues. That set
of assumptions includes those mentioned above, as well as the assumptions that the time between arrivals and the time of service
are both represented statistically by the exponential distribution.
In fact, as we show later, some of these assumptions are not critical.

The classical approach proceeds by defining a state space for the network of stations, or FMS in our viewpoint. Let each state for an FMS be the vector representing the current number of parts at each station [including the part(s) being processed]. In particular, notice that this notion of state does not explicitly consider the amount of time that remains for any parts being processed, nor does it represent the relative ranking of parts in the station queue. These kinds of information are, however, of fundamental importance to simulation modeling.

The objective of the classical approach is to derive the equilibrium probability of the FMS being in a particular state. A detailed development of that equilibrium probability solution is shown in the appendix to this chapter. This solution is known as the "product-form" solution for the FMS. It contains a normalizing constant $G(M,N)$, where M is the number of stations in the FMS and N is the number of parts circulating in the closed system. If M and N are varied, a matrix of values for such constants can be associated with the FMS. As we see below, this matrix plays a critical role in the use of n-o-q models to calculate FMS performance measures.

The product-form closed-network solution yields one of the n-o-q analysis methods, namely CAN-Q [2]. Prior to describing the details of that method, let's pause to interpret the application of this approach. How does a closed queueing network relate to the behavior of an FMS? If we assume that an FMS has N parts being processed in it, and that as soon as one part leaves, it is replaced by another, we can interpret the FMS as having N parts which recirculate indefinitely. Thus, the notion of a closed network is not such a poor approximation from this perspective. However, it should be noted that the closed-network results derived in the chapter appendix also assume that the service time at each station is identical for all servers *and* for all parts. This can be interpreted as a system that has only *one* part type being produced and cannot have different resources at each station (e.g., a mix of old, slow machines with new, faster machines).

Theoretical results have been derived for closed networks with more than one part class. Such results, however, apparently are not as amenable to computational methods as for the one-class case. In order to obtain the ease of computation provided by the one-class case (described in detail below), the restriction on variety of part classes could be (and, in fact, is) surmounted by defining, for those cases where there are multiple part types, a kind of "aggregate part type" which represents all the individual part types on an *average* basis.

The closed network n-o-q analysis method, as manifest in CAN-Q [3], is based on Buzen's algorithm (described in this chapter's appendix) and proceeds by calculating the elements of a matrix where the ijth element is of the form $G(i,j)$. Here, the row component i refers to a number of stations, and the column component j refers to the number of parts in the system. The total number of stations in the system is taken to be M and the total parts to be N. In particular, the ratio

$$G(M, N - 1)/G(M,N)$$

which would be obtained when the M-by-N matrix is filled out, is particularly meaningful. Solberg showed that this ratio can be used to calculate several important performance measures for n-o-q's. To do so, he envisioned an FMS as a set of processing devices (machines) and an mhs for moving parts between processors. In fact, this mhs is ubiquitous since parts can only get from one machine to another via the mhs. Thus, he arrived at a schematic for FMS that appears as shown in Fig. 9.4.

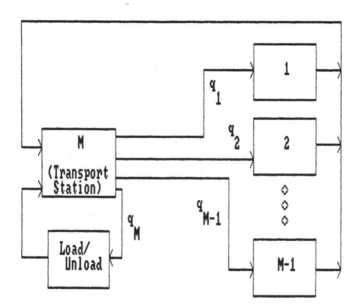

Fig. 9.4 Generic CAN-Q model for FMS.

Since parts arrive as fast as they are completed (to maintain N parts circulating in the system), parts are represented as leaving the system by their probability q_M of going to the unload station. Since all parts move through the mhs, the average delivery rate per transporter, w_M, should clearly be an important factor in any performance measure for the system. This is shown by the following formulas that Solberg derived for the performance measures of system production rate and production time.

The steady-state average number of completed parts per time unit (e.g., per hour) is called the system production rate and is given by

$$P = q_M \times w_M \times G(M,N - 1)/G(M,N)$$

As explained below, q_M and w_M are data provided by the CAN-Q user, and so P can be computed as soon as the G matrix is computed.

The average time spent by a part in the FMS system, T, is given by

$$T = N/(q_M \times w_M) \times G(M,N)/G(M,N - 1)$$

and again we note that since N is user-specified data, T can be calculated when the G matrix is known. The value of T not only represents the processing and handling time of parts in the system, but also includes the time that those parts spend waiting for processing or movement. In this manner, the effects of congestion are included in the system performance measures.

In addition to the two important performance measures given above, we are also interested in knowing the utilizations of each of the processing stations as well as the delay that parts incur at each station. The utilization of station i may be computed as

$$u_i = (q_i \times w_M/w_i) \times G(M,N - 1)/G(M,N)$$

where q_i is derived from the user-specified data and refers to the probability of a part being routed from the mhs to station i. The value of w_i, the work rate of station i, is also obtained via data supplied by the user. Finally, for stations that have only one server, the average number of parts at those stations, including the one in service, is denoted $E(x_i)$.

$$E(x_i) = \sum_{k=1}^{N} (r_i)^k \times G(M, N - k)/G(M, N)$$

where $r_i = (q_i \times w_M/w_i)$, the multiplier for u_i above. Then, for station i, the expected number in queue is

$$E(x_i) - u_i$$

As noted earlier, the closed-network approach is implemented in CAN-Q for the single-part-type (for both single and multiple servers) case. Yet, almost by the definition of FMS, these manufacturing systems are intended for production of at least several part types simultaneously. Therefore, the input data for the FMS that makes multiple part types must be transformed in order to apply this closed-network n-o-q method. This transformation is done in a manner similar to the calculations we did at the beginning of this chapter when we tried to establish a "bridge" between simulation and n-o-q.

Let us consider the n-o-q model parameters that are needed and show how they might be derived from real FMS data supplied by the user. Consider first the parameter q_i, which is defined as the probability of routing any part to station i. To estimate this, picture an FMS that makes two part types on three machines and one inspection station, and imagine for the moment only one part type, say product A, in the FMS. All parts of product A visit three machines, and every second such part visits a fourth inspection station. If we observe the parts coming off the mhs over a long time, we would expect that a proportion of about 1/3.5 of them goes to each of the three machines that all parts visit, and a proportion of about 0.5/3.5 of them goes to the inspection machine. Let us now introduce the second product type B to the FMS and arrange the mix of these two types so that there are an equal number of parts of each type in the system. All parts of type B visit two of the machines, which are also visited by all products of type A. Observation of this new situation in the FMS over a long time would be expected to show that a proportion of 1/5.5 of the parts coming off the mhs visits one of the machines (the one that works on type A only), and that about 2/5.5 of the parts visit each of the other two machines, and finally that 0.5/5.5 of the parts go to the inspection station. These proportions are in fact the branch probabilities, q_i, that are used by CAN-Q. In the more general case where the part mix is not equally divided, the part mix ratio alters these proportions by essentially weighting the terms in the proportions according to the ratio. It is important to note that we have "aggregated" the two product types into *one* "aggregate-product" type, in order to achieve efficiency of computation of the system performance measures.

CAN-Q also aggregates the operation times at each station into *one* operation time for the corresponding "aggregate-product" type. This is done in a manner similar to above by using visit frequencies and part mix information supplied by the user. For example, for the equally divided part mix case, if product A has one operation on a machine of duration 90 minutes and if product B has one operation on the same machine with duration 30 minutes, and if all parts of type A visit the machine but only half the type B parts do, then an aggregate operation time for the machine is calculated as follows:

$$1/w_i = (90 + [0.5 \times 30])/(1 + 0.5) = 105/1.5 = 70 \text{ minutes}$$

Once the (possibly) aggregated parameter values are calculated from the input data, the CAN-Q program is able to compute the G-matrix and the system performance measures described earlier.

9.3.1 Example of n-o-q Analysis

This example is adapted from the one in the *CAN-Q User's Guide* [2]. Consider a small system of three stations producing two product types. The stations include a milling station having two mills, a drilling station with three drills, an inspection station, and two forklift trucks for material handling. The layout for the system is pictured in Fig. 9.5. Table 9.1 displays the data needed to model this system. Assume that the product mix is equally distributed between part types A and B. Also assume that the average time to move a part from one station to another is 4 minutes.

The input data for CAN-Q is shown in Table 9.2. The first two rows of data give output information. The third row indicates how many stations, parts (i.e., the value of N), and product types are to be in the FMS. A rule of thumb for selecting N is used here; namely, "make N equal to three times the number of servers." The stations (including transport and transport time) are described on the next four lines. Finally, the two product types and their process plans are given.

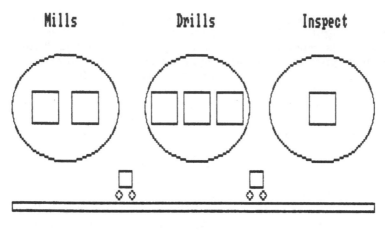

Mills   **Drills**   **Inspect**

Transport

Fig. 9.5 Layout of CAN-Q FMS example.

The output of the CAN-Q analysis is shown in Table 9.3. The overall production rate as well as the rate by part type are given. The average time that the parts spend in the system is shown in an aggregate form, and also in terms of its constituent times. The next part of the output gives information regarding the value of N. Note, for example, that reducing N to 15 would barely decrease production rate, but would reduce time in the system by about an hour. Maximum production rate occurs at N = inf.

Finally, the CAN-Q output identifies the bottleneck station. This is the station that has the heaviest workload. For this station and all the others, CAN-Q provides utilization and waiting time data (as shown for station 1).

The full CAN-Q output also gives sensitivity information regarding changes in processing time at stations as well as changes in product mix. The importance of this sensitivity information will be discussed in the next chapter.

9.4 MEAN-VALUE ANALYSIS: A SECOND n-o-q APPROACH

The classical Jackson-network approach discussed in the previous section has been implemented in algorithmic form as, for example,

Table 9.1 Data for CAN-Q FMS Example

	Facility	
Station	Name	Number of servers
1	Mill	2
2	Drill	3
3	Inspect	1
4	Transport	2

		Process plan A	
Operation	Station	Duration	Frequency
1	1	20.0	1
2	2	30.0	1
3	3	15.0	.5
4	1	14.0	1

		Process plan B	
Operation	Station	Duration	Frequency
1	2	90.0	1
2	1	30.0	1

Table 9.2 CAN-Q Input Data
for FMS Example

```
CANQ User's Guide example
0031
      3    18    2
Trans    2    4.0
Mill     2
Drill    3
Insp     1
Proda    4    .5
      1   20.
      2   30.
      3   15.
      1   14.
Prodb    2    .5
      2   90.
      1   30.
```

Table 9.3 CAN-Q Analysis Output

System performance measures

Production rate = 2.978 items/hour

Production rates by product type

	Number	Value
Proda	1.489	1.489
Prodb	1.489	1.489
Total value =		2.978

Average time in system = 362.71 minutes

Processing	99.50
Traveling	12.00
Waiting	251.21

Functions of N, number of items in system

N	Production rate	Average time in system
1	0.538	111.500
.	.	.
.	.	.
.	.	.
13	2.927	266.460
14	2.943	285.432
15	2.955	304.568
16	2.964	323.840
17	2.972	343.225
18	2.978	362.706
19	2.982	382.268
20	2.986	401.899
21	2.989	421.587
22	2.991	441.325
23	2.993	461.105
.	.	.
.	.	.
,	.	.
inf	3.000	inf

The bottleneck station is 2

(continued)

Table 9.3 (Continued)

Summary for station number 1 : Mill

No. of servers	Server utilization	Ave. no. of busy servers
2	0.794	1.588

Steady-state average number of

Items at station	3.980
Items in process	1.588
Items waiting	2.392

Average time spent at this station	Per operation	Per item
Total time (minutes)	53.462	80.193
Processing	21.333	32.000
Waiting	32.129	48.193

Fraction of time	× Items at station	× Items exceeded
$x = 0$	.1144	.8856
$x = 1$	.1831	.7025
$x = 2$	.1465	.5560
$x = 3$	.1172	.4388
$x = 4$	.0937	.3451
$x = 5$	.0750	.2701
$x = 6$	.0600	.2101
$x = 7$	.0480	.1621
$x = 8$	.0384	.1238
$x = 9$	.0307	.0931
$x = 10$	.0245	.0686
$x = 11$	.0195	.0490
$x = 12$	.0155	.0335
$x = 13$	.0121	.0214
$x = 14$	.0092	.0122
$x = 15$	.0065	.0057
$x = 16$	.0038	.0018
$x = 17$	.0015	.0003
$x = 18$	.0003	.0000

the CAN-Q program. A limitation of the classical approach is the
difficulty of dealing with multiple part types. A second approach
to n-o-q models, called mean-value analysis (MVA), alleviates this
problem. As its name implies, MVA uses only information about
the *mean* value of system parameters rather than the entire statistical distribution which is assumed for the classical approach.

The basic relationships for MVA are quite intuitive and, again,
based on closed-network assumptions and long-term average behavior for the modeled system. The fundamental relation, as expressed by Reiser and Lavenberg [4], expresses the mean waiting time (including service time) for a given part type, c, at a
(single server) station, m, in terms of the mean service time for
that part type and the mean queue size at that station:

$$\tau_{c,m} = t_{c,m} + t_{c,m} \times q_m^{-1}$$

where $\tau_{c,m}$ is the mean waiting time (including service) for part
type c at station m, $t_{c,m}$ is the mean service time of part type c
at station m (NOTE: for now, we assume that $t_{c,m} = t_m$ for all
part types c; that is, all part types have the *same* service time
on station m; this restrictive assumption will be removed) and
q_m^{-1} is the mean queue size at station m WITH ONE LESS JOB
OF PART TYPE c IN THE SYSTEM.

The q_m^{-1} term looks peculiar as stated, but can be viewed in
a perhaps more revealing light as stating that, in steady state,
an arriving part sees the station as if it (the part) was *not* in
the system. Stated differently, the mean time a part stays at a
station equals its own mean service time plus the mean backlog
on arrival [i.e., the mean backlog being the average queue length
at the station, if it (the part) were not in the system, multiplied
by the average service time of those parts at that station].

The notion that average waiting time for a part depends on
backlogs calculated as if the part were not in the system forms
the basis for an (iterative) algorithm that calculates performance
measures for n-o-q models. The algorithm begins by assuming
that the system is empty and proceeds to add one part of each
type at a time, continuing to update the mean waiting times and
queue lengths, until the specified number of each part type is
present (recall that this approach is a closed-network one, so
the number of each part type in the system is prespecified by
the user).

The MVA iteration scheme calculates queue lengths using two
relationships developed in queueing theory more than two decades

ago [5] in order to obtain system performance measures. In particular, Little's formula for a single station states that mean number of parts of type c at station m is as follows:

$$q_{c,m} = \lambda_c \times \tau_{c,m}$$

where λ_c is the throughput for part type c. λ_c is calculated as follows:

$$\lambda_c = N_c / \sum_m \tau_{c,m}$$

where N_c is the number of parts of type c in the (closed) system.

An iterative procedure can now be presented for obtaining the throughput and queue lengths of FMS represented by mean value n-o-q models:

1. Initialize all mean queue lengths, q_m, to zero.
2. Starting with an empty system, introduce the N_1 parts of type 1 *one* at a time, each time calculating $\tau_{c,m}$, λ_c, and q_m for all part classes and stations. (After the first iteration, the value of $\tau_{c,m}$ is calculated from the value of q_m obtained on the previous iteration.)

$$\tau_{c,m} = \ldots$$

$$\lambda_c = \ldots$$

$$q_m = \sum_c q_{c,m} = \ldots$$

3. Now introduce one part of type 2 and reintroduce parts of type 1 one at a time, each time calculating as in step 2.
4. Continue for all parts of each type: N_1 of type 1, N_2 of type 2, etc.
5. Iteration stops when all parts of every type have been introduced into the system.

The throughput of the system for each part type as well as the mean waiting times and mean queue lengths have now been found for the n-o-q model.

9.4.1 MVA Example

Suppose that an FMS is proposed that is to produce two part types with the following operation times on four machines:

	Part type A	Part type B
Machine 1	90	90
Machine 2	25	25
Machine 3	50	
Machine 4		10

Assume that the material-handling time is negligible with respect to machining times. We can illustrate the mean value iteration as follows: Starting with an empty system, add one part of type B. Since, without that part, the queue lengths must have been zero, the mean waiting time values are

$$\tau_{A,1} = 90 \qquad \tau_{A,2} = 25 \qquad \tau_{A,3} = 50$$

$$\tau_{B,1} = 90 \qquad \tau_{B,2} = 25 \qquad \tau_{B,4} = 10$$

The throughputs for the part types are

$$\lambda_A = 0 \qquad \lambda_B = 1/(90 + 25 + 10) = .008 \text{ parts/minute}$$
$$= 3.84 \text{ parts/hour}$$

The new mean queue lengths, q_m, at each station become

$$q_1 = q_{A,1} + q_{B,1} = 0 \times 90 + .008 \times 90 = .72$$

and, similarly,

$$q_2 = .2 \qquad q_3 = 0 \qquad q_4 = .08$$

Now another part of type B is added to the system and the performance measures are recalculated, continuing to do so up to the fixed number of parts of type B in the closed system.

At that point, a single part of type A is added to the system and the above iteration is performed for *all* of the parts of type B. This "outer loop" continues until all parts of both types have been entered into the FMS. The resulting values of λ_A and λ_B

give the throughputs for each part type, and the q_m values provide the mean queue lengths at each station.

The above scheme for MVA can be quite inefficient in terms of computation time for systems with moderately large numbers of stations and part classes. Reiser and Lavenberg suggested a modification to the basic approach that improves the computational efficiency and extends its generality, at the cost of obtaining only an approximate solution. Part of the change consists of allowing the service times at each station to be *different* for each part class. This results in an equation for $\tau_{c,m}$ of the form

$$\tau_{c,m} = \tau_{c,m} + \sum_j t_{j,m} \times q_{j,m}^{-1}$$

The new term $t_{j,m}$, represents the potentially different service times for each part class j on station m. Though this term enhances the generality of the algorithm, it is the suggested change in the form of $q_{j,m}^{-1}$ that improves the computational characteristics. In particular, R&L approximate $q_{j,m}^{-1}$ by the sum

$$q_{j,m} + \varepsilon_{c,m}j$$

where the $q_{j,m}$ term is *no longer recursively* dependent on one less part in the system, and where the ε "correction term" is used instead to correct for one less part of type j. This form for $q_{j,m}^{-1}$ has the effect of replacing the multiple-loop recursion in the basic MVA scheme shown above with a simple iterative algorithm as follows:

1. $\varepsilon_{c,m}j = \ldots$
2. $\tau_{c,m} = \ldots$
3. $\lambda_c = \ldots$
4. $q_{c,m} = \ldots$
5. back to 1

Iteration is continued until some convergence criterion is met. For example, one such criterion could be to continue until the largest difference between $q_{c,m}$ in consecutive iterations falls below some prespecified value. Though this iterative scheme has been shown analytically to converge for only a few simple cases, it often converges quickly in practice.

This second, more general MVA scheme has been used to calculate n-o-q models for FMS systems, especially by the researchers at Draper Labs [6]. Their algorithm, called MVAQ, is an implementation of an extension of the Reiser/Lavenberg approach made by Schweitzer [7] and Bard [8]. The extension results in a modification to the recursive $q_{c,m}^{-1}$ term to obtain simple iteration.

Comparisons of the MVAQ algorithm have been made to simulation models of equivalent FMS systems. For example, a system with four machines, six pallets, and a conveyor mhs that processed 10 different part types was modeled using both simulation and the MVAQ n-o-q model. The results showed that although the queue sizes were very close in comparison, the MVAQ model consistently produced estimates of throughput and machine utilization that were about 10% low of the simulation values [9].

Any comparisons of MVA algorithms to simulation results must, however, be accompanied by the caveat that the MVA programs do *not* account for "blocking" in the system. "Blocking" refers to impediments to movement of parts through the system. The MVA approach assumes that a part completing service at one station can immediately move on to the next station and allow the next part to be served. In real FMS systems, however, it is possible for a part to be blocked from moving into its next machine because of material-handling congestion or breakdown, machine breakdown, and so forth. Therefore, if the simulation model (and, of course, the actual system) displays such blocking behavior, the MVA results must be viewed as suspect. Of course, the same caveat is also applicable to n-o-q modeling based on the classical approach, as discussed in the previous section.

9.5 OPERATIONAL ANALYSIS APPROACH TO n-o-q MODELS

Suri [10] points out that operational analysis (OpA) was advanced by Buzen [11] and extended by Denning and Buzen [12] in order to explain the surprising applicability of n-o-q models in the face of their seemingly restrictive assumptions. Suri's article gives further justification for the robustness of the n-o-q methods.

The underlying philosophy of OpA is interesting in that one of its basic principles states: "All quantities (used in the analysis) should be defined so as to be *precisely measurable*, and all assumptions should be stated as to be *directly testable*."

The OpA approach is based on four assumptions which involve *no* statement of the underlying probability distribution of the system's service times. The assumptions may be stated in terms of the state of the system where, again, the state of the system at any time is defined by the number of parts at each station. The sum of the parts at each station over all the stations is N (i.e., a closed system). Note that this is the same definition of system state as for other n-o-q methods.

The four assumptions of OpA are

1. During the time that the system behavior is observed, the number of transitions to any state is equal to the number of transitions out of that state.
2. State changes result only from single parts moving between pairs of stations in the system.
3. The output rate of a station is determined completely by its queue length and is otherwise independent of the system's state.
4. The routing frequencies are independent of the system's state.

Assumption 1 is, by now, the usual one made for n-o-q analysis. The second assumption implies that no *simultaneous* part moves will be observed in the FMS. This gives rise to the same type of state transitions as were postulated for the Williams and Bhandiwad development in the chapter appendix. Assumption 3 refers to the restriction that no station can *block* any other one. This excludes the possibility of keeping a part at a station because the waiting area at its next station is full. Finally, the fourth assumption indicates that part movement is not a function of instantaneous queue lengths in the system. So it might not be reasonable to expect accuracy from applying OpA methods to manufacturing systems in which parts "*balk*" (i.e., are rerouted) from a station because its waiting area is full.

Note that, as for the classical approach, our attention is focused on systems with only *one* part type. Again, the interpretation is made that one of the stations is a "load/unload" station and that, as soon as a part leaves (i.e., is "unloaded") from the system via this station, another is introduced in its place.

With the above assumptions, equations can be derived for system performance measures such as throughput and station utilizations which are similar in form to the equations derived using the classical approach (that is, they too are functions of the normalization constant G calculated for both (N - 1) and N parts in the

system). These equations, however, also depend on a term that
Suri contends can only be determined via detailed simulation, thus
defeating the purpose of the less detailed n-o-q models. In order
to remove this dependence and allow calculations to proceed with
only the type of data previously described as input to CAN-Q, a
fifth assumption is added to OpA;

5.   The mean time between completions at any station is indepen-
     dent of the number of parts at that station.

Given the five assumptions, relatively simple equations for system
performance measures can be derived. Suri's article shows that
assumption 5 need not be viewed as being restrictive since the
OpA calculations are quite robust to violations of that assumption.

     Some controversy has developed over the difference between
the OpA approach and that of the classical approach. In a forum
discussion [13], Bard and Muntz have contended that the assumptions of OpA are essentially the same as their classical counterparts, whereas Buzen has argued otherwise. For our purposes
in this book, it is only important that both approaches have been
developed to the point that useful algorithms for use by FMS designers have resulted from the researcher's efforts.

9.6 SUMMARY

The three main types of n-o-q models useful for FMS analysis and
design have been presented in this chapter. The characteristics
of the three types are summarized in Fig. 9.6. Relative to simulation, all the models are useful in predicting system behavior at
less cost of data collection and computer time. Furthermore, the
predictions can be surprisingly accurate. Given that "there is
no free lunch," these n-o-q methods also suffer from some draw-
backs. As FMS design proceeds, concern is centered on more detailed design decisions. Behavior of the system that may be crucial to these decisions may include, say, blocking of material movement. At this point, the use of n-o-q methods may no longer be
appropriate.

     Within the set of three methods, there are some relative advan-
tages. For example, the mean-value approach can more easily accommodate a large number of different classes of parts within the
system. On the other hand, the classical and the OpA approaches
provide more information, giving not only the mean values of

Approach	Characteristics
Classical Approach and	Provides distributions of significant performance measures, as well as mean values
Operational Analysis	Single-part-type analysis more tractable than for multiple part types
	Sensitivity data are calculated directly
Mean value Analysis	Consideration of multiple part types is more tractable
	Provides only mean values of performance measures
	Does not directly provide sensitivity data

Fig. 9.6 Characteristics of n-o-q models.

performance measures, but in fact the entire distribution for the length of queues. The latter data provide the designer with the ability to assess not only the mean queue length at a station, but also the frequency with which that queue exceeded a certain length. Such information is useful for buffer sizing.

Another advantage of the classical and OpA approaches is their provision for *analytical* derivation of sensitivity measures for the performance measures with respect to the parameters of the system. This property may be exploited as described in the next chapter in order to aid the FMS designer.

APPENDIX: PRODUCT-FORM SOLUTION
FOR CLASSICAL APPROACH

The objective of the classical approach is to derive the equilibrium probability of the FMS being in a particular state. For example, the probability of a zero-state vector gives the likelihood that the system is empty, and therefore that all servers are idle. It is important to note that the equilibrium probability applies to the system behavior only when the system is in "steady state." This steady-state assumption is an inherent assumption of all n-o-q methods and must be kept in mind when the results of an n-o-q analysis are to be interpreted. For practical purposes, the notion of steady state implies that a person observing the system would see the same subset of states recurring at various times. So, for example, if an FMS started with all stations empty and if the average arrival rate was high, the state represented by the zero-state vector might be observed only a few times at the beginning of observation. Later, this state may not be seen again. Thus, this system was not initially in steady state.

As noted above, we are interested in expressing the probability of occurrence of system states in a steady-state context in terms of the behavior of the individual queues. We shall limit our consideration to stations with single servers and to *closed* systems only (i.e., systems in which, with respect to Fig. 9.3, no new parts come IN and no parts in the system DEPART; that is, consider systems in which N parts appear to circulate). We begin to do so by expressing the fundamental relationship that holds for each individual station as depicted above: With the state denoted by $(n_1, n_2, \ldots, n_M)$, where n_i is the number of parts at station i (including the one being processed), we can interpret equilibrium as follows: the rate of transitions out of the state $(n_1, n_2, \ldots, n_M)$ is *equal* to the rate of transitions into this state.

Transition from a state implies that a part leaves a station j to enter station k. The mean service time at station j is denoted τ_j. Then (owing to the assumption of exponential service times), the probability of leaving station j in time interval δt is $\delta t / \tau_j$. The part will then visit station k with probability p_{jk}. In equilibrium, for any state S the probability of being in that state, $\Pr(S)$, times the probability of a transition from that state has to be equal to the sum over all states S' of $\Pr(S')$ times the probability of a transition from S' to S. Then, as in Williams and Bhandiwad [14], referred to below as W&B, we can write (dropping δt's),

$$\sum_{j \,\mid\, n_j > 0} (1/\tau_j)\, \Pr(n_1, n_2, \ldots, n_M) =$$

$$\sum_{j \,\mid\, n_j > 0} \sum_i (p_{ij}/\tau_i)\, \Pr(n_1, \ldots, n_i + 1, \ldots, n_j - 1, \ldots, n_M)$$

which mathematically expresses the above equilibrium relationship.

This equation has the solution

$$Pr(n_1, n_2, \ldots, n_M) = (1/C) \prod_{j=1}^{M} X_j^{n_j}$$

where $X_j = \tau_j \times \rho_j$. ρ_j may be thought of as the "relative demand" for station j. That is, the relative demand for station j is composed of the relative demands for all stations i that send parts to j, weighted by the fraction of those which go from i to j. So,

$$\rho_j = \sum_i \rho_i \times P_{ij}$$

Solberg [3] shows that, in an FMS, X_i can be interpreted, for each station i, as the utilization at station i relative to the utilization of the material-handling system.

The equilibrium equation solution is referred to as the product-form solution for the network under equilibrium conditions. The constant C is determined by the requirement that the sum of probabilities over all states is unity. So,

$$C = \sum \prod_{i=1}^{M} X_1^{n_i}$$

where the sum is over all $(n_1, n_2, \ldots, n_M)$ such that $\Sigma n_i = N$. Recall that we are considering the closed-system case where N parts circulate in the system.

The restriction that $\Sigma n_i = N$ is important to W&B's development because it motivates them to consider the following polynomial:

$$g(t) = (1 + X_1 t + X_1^2 t^2 + \ldots) \ (1 + X_2 t + X_2^2 t^2 + \ldots) \ \ldots$$

$$(1 + X_M t + X_M^2 t^2 + \ldots)$$

Examination of the expansion of g(t) reveals that the coefficient of t^N is the same as the normalizing constant C. That is, the coefficient of t^N is the sum of terms of the form

$$X_1^{n_1} X_2^{n_2} \ldots X_M^{n_M}$$

with one such term for *each possible way* of choosing the n_i such that $\Sigma n_i = N$.

Notice that if the first term, 1, is left out of the first factor of $g(t)$, then the coefficient of t^N will *not* contain any terms of the form

$$X_1^{\ 0} \ X_2^{\ n_2} \ldots X_M^{\ n_M}$$

In fact, the coefficient of t^N will be the sum of all terms for which $n_1 \geqslant 1$. Since our notion of the queue length includes the part being processed, this coefficient represents all possibilities for which station 1 is *busy*. Thus, the coefficient of t^N in the modified $g(t)$ divided by the corresponding coefficient in the original $g(t)$ (which includes zero length queues at station 1) will provide an important performance criterion value, namely, the *utilization* of station 1.

W&B refer to $g(t)$ as the generating function for the network. They express the expansion of $g(t)$ (for M stations) as

$$g(t) = 1 + G(M,1)t + G(M,2)t^2 + \ldots + G(M,N)t^N + \ldots$$

By manipulating the factors of $g(t)$ as above, they quickly show that the utilization of any station can be expressed as

$$U_i = Pr(n_i \geqslant 1) = \frac{X_i G(M,N-1)}{G(M,N)}$$

Furthermore, the expected queue length at any station i is given by

$$Q_i = [1/G(M,N)] \sum_{j=1}^{N} X_i^{\ j} G(M,N-j)$$

Other performance measures can also be derived in terms of the $G(M,i)$.

The efficiency of n-o-q methods results from the fact that the G values can be calculated in a recursive manner. Consider the original definition of $g(t)$ above. Denote the first (leftmost) factor of $g(t)$, containing only the X_1 terms, as $g_1(t)$. Consistent with the previous interpretation, the coefficient of t^j in $g_1(t)$

represents all the ways in which j parts can be allocated at the single station being considered in this factor, namely all at station 1. If we extend our consideration to two stations, i.e., to the first two factors of g(t), then the coefficient of tj in this expansion, denoted $g_2(t)$, for these two factors shows all the ways that j parts can be allocated among the two stations. Finally, extend the definition to $g_i(t)$:

$$g_i(t) = g_{i-1}(t) \, x_i(t)$$

where $x_i(t)$ is the ith factor of g(t), in which case $g_N(t) = g(t)$. Then, if we define G(i,j) to be the coefficient of t^j in $g_i(t)$, the following recursive equation is easily shown:

$$G(i,j) = G(i-1,j) + X_i G(i,j-1)$$

This algorithm is given in Buzen's paper [11].

REFERENCES

1. J. R. Jackson, "Jobshop-like queueing systems," *Journal of TIMS, 10*: 131 (1963).

2. J. Solberg, *CAN-Q User's Guide*, report no. 9, NSF Grant No. APR74 15256, Purdue University (1980).

3. J. Solberg, "Optimal Design and Control of Computerized Manufacturing Systems," Proc. of AIIE Systems Engineering Conf., Boston, pp. 137–147 (1976).

4. M. Reiser and S. S. Lavenberg, "Mean-value analysis of closed multichain queuening networks," *Journal of ACM, 27*: 313 (1980).

5. J. C. D. Little, "A proof of the queueing formula L-lambda W," *Operations Research, 9*: 383 (1961).

6. R. Suri and R. R. Hildebrant, "Using mean value analysis," *Journal of Manufacturing Systems, 3* (1984).

7. P. Schweitzer, "Approximate Analysis of Multiclass Closed Networks of Queues," Int'l. Conf. on Stochastic Control and Optimization, Amsterdam (1979).

8. Y. Bard, "Some extensions to multiclass queueing network analysis," in *Performance of Computer Systems* (M. Arato, ed.), North Holland, Amsterdam (1979).

9. R. R. Hildebrant, "Scheduling Flexible Machining Systems Using Mean Value Analysis," Proc. of IEEE Conf. on Decision and Control, Alberquerque (1980).

10. R. Suri, "Robustness of queuing network formulas," *Journal of ACM, 30*: 564 (1983).

11. J. P. Buzen, "Computational algorithms for closed queueing networks with exponential servers," *Comm. ACM 16* (1973).

12. P. J. Denning and J. P. Buzen, "The operational analysis of queueing network models," *ACM Comput. Surv., 10*: 225 (1978).

13. "Surveyor's forum," *ACM Comput. Surv., 11*: 69 (1979).

14. A. C. Williams and R. A. Bhandiwad, "A generating function approach to queueing network analysis of multiprogrammed computers," *Networks, 6*: 1 (1976).

10

Network-of-Queue Analysis as a Design Aid for FMS

10.1 INTRODUCTION

Our purpose in this chapter is develop the foundations of an automated design environment for n-o-q modeling aids to the FMS designer. To begin, we shall consider a rough-cut approach to FMS design which could be used to structure the very preliminary data initially available to the designer. The aim of this structure is to arrive at a set of data that can be input to n-o-q models for more accurate performance estimation.

10.2 "ROUGH-CUT" FMS DESIGN

It is difficult to imagine designing an FMS without having information available on part types to be made and the processes needed to make them. So we shall assume that at least that much information is available. (Methodology discussed in Chapter 12 will show one way to select part types to be made on an FMS.) In particular, let us suppose that at least one sequence of operations along with the operation times is specified for each part type. The sequencing information may also include requirements for

operations to be performed on different machines, and for one
or more fixturing setups, in which case we would need some data
on fixturing times. Furthermore, let us ask that material-handling (mh) information be available, at least to the extent of
estimating the speed of mh devices. Finally, production requirements and resource availability need to be stated in order to estimate the size of the FMS.

An illustration of this rough-cut approach can be given with
the following part-type data:

Operation Times in Minutes (Hours) for Two Part Types

Operation	#1	#2
10	17.4 (.29)	16.2 (.27)
20	3.6 (.06)	3.6 (.06)
30	9.0 (.15)	9.0 (.15)
40	22.8 (.38)	28.2 (.47)
50	3.0 (.05)	3.6 (.06)
60	14.4 (.24)	19.8 (.33)
70	13.2 (.22)	21.6 (.36)
Cycle time	83.4 (1.39)	102.0 (1.70)

Assume that five different processes are associated with the two
part types. In particular, let operations 30 and 40 be performed
via identical processes and assume the same is true for operations
60 and 70. No specification is given as to whether such processing must be performed on specialized machines or on general-purpose four- or five-axis machining centers.

Let us say that the production requirements for these part
types are set at 1500 parts/month (equally divided between the
two types), and that the system will be assumed to be in operation for 250 hours/month.

10.2.1 Estimation of Number of Machines

The first step in the rough-cut design procedure will be to estimate the number of machines to be in the FMS. This can be done
under either of two "idealizing" assumptions:

1. All operations are done on one station, and as soon as one
    part is completed, the next is loaded for processing (and so
    each station is utilized 100% of the time).

2. Operations are done on different stations, each station pos-
    sibly having more than one machine, with each station being
    kept 100% utilized.

Each assumption can be justified under different conditions, and
each leads to a procedure for estimation of number of machines.
Assumption 1 would seem to be most applicable when the FMS is
envisioned to be composed of general-purpose machining centers,
whereas assumption 2 appears to be more appropriate for the case
of specialized machining stations.

If we use assumption 1, we can estimate the number of stations
as follows:

Number of stations = (required production rate) ×
(part type cycle time)

Here, the required production rate is 1500/250, or 6 parts/hour.
The average part type cycle time is 1.55 hours; so the number of
stations needed for the FMS is estimated by this approach to be
10 (9.3 rounded up to the next integer).

The use of assumption 2 requires a little more thought and a
bit more calculation. Imagine each operation to be done on a separate station, unless consecutive operations are specified to be
done via the same process. For our example, this leads to a consideration of five stations, one for operation 10, one for 20, one
for 30 and 40, one for 50, and one for 60 and 70. (If a part type
returns to a previously used process after intervening ones, the
second usage is treated as yet again a separate station.) In effect, we have created a "pipeline" through which parts flow
from raw to finished state. The main limitations on the flow of
such a pipeline are "restrictions" in the pipeline itself. Our analog to such restrictions is long operation times. If long operation
times can in effect be reduced, the parts will flow faster from the
pipeline. One way to, in effect, reduce such operation times is
to provide more servers at each station.

In this example, the longest operation time on a given station
occurs for operations 60 and 70 of part type 2 and is equal to .69
hour. So even if all stations are utilized to their maximum extent,
no two parts of type 2 can be completed with less than .69 hour
between their departure times. Thus these parts cannot be produced at a faster rate than 1/.69 or 1.45 parts/hour, below our
desired production rate of 3 parts/hour for each type (recall that
we specified equal quantities of each type). How many machines

do we need at this station? Using the formula given for assumption 1 (which applies for parallel servers) and noting that with an equal part mix the desired production rate for each type is 3 parts/hr, we get

Number of machines = 3 × .46 + 3 × .69 = 3.45, rounded up to 4

Proceeding in a similar manner, we estimate the number of machines at all stations as

Station 1 (op 10)	1.68 machines, rounded up to 2
Station 2 (op 20)	.36 machine, rounded up to 1
Station 3 (ops 30, 40)	3.45 machines, rounded up to 4
Station 4 (op 50)	.33 machine, rounded up to 1
Station 5 (ops 60, 70)	3.45 machines, rounded up to 4

This "pipeline" approach to sizing the system gives a more conservative estimate of 12 machines in the system. It also provides additional information on the relative number of different (presumably specialized) machines in the system.

10.2.2 Estimation for Material-Handling Stations

Material handling in the FMS includes fixturing/defixturing (f/d) stations, also referred to as load/unload stations, as well as the part transportation system. Given a specified number of fixturings for each part type, we can estimate the maximum time allowable for the fixturing process. This information in turn can be used to estimate the number of servers at that station. The number of fixturings also relates to part transport since fixtured parts must at least be carried to stations and returned for unloading.

For our example, suppose that the parts will require three fixturings, the first of which will carry the parts to operations 10 and 20 (e.g., for the machining of locator pads to be used in subsequent operations). Let there be a second setup for operations 30, 40, and 50. The parts are finally fixtured a third time for the remaining operations.

Fixturing may affect the calculation of system size via assumption 1 above. In particular, we had previously assumed that any part could visit any station and receive all its operations (by the one server) there. With fixturing, each part visits three different stations in our example, one station for each fixture type.

The sizing calculations under assumption 1 would use the following data:

Operations	#1	#2	Avg.
Station 1 (ops 10, 20)	21.0 (.35)	19.8 (.33)	(.34)
Station 2 (ops 30, 40, 50)	34.8 (.58)	40.8 (.68)	(.63)
Station 3 (ops 60, 70)	27.6 (.46)	41.4 (.69)	(.58)

The required production rate is still 6 parts/hour, and since every part visits these three stations, we can calculate the number of servers required at each station as follows:

Station 1 $6 \times .34 = 2.04$, which rounds up to 3
Station 2 $6 \times .63 = 3.78$, which rounds up to 4
Station 3 $6 \times .58 = 3.48$, which rounds up to 4

The estimated system size is now 11 machines to be viewed as partitioned into three stations.

Given a desired production rate of 6 parts/hour and a need for three setups per part, the f/d station must perform at about 36 f/d operations/hour. This leads to an estimate of a maximum f/d time of 1/36, or .028 hour. This period of a little under 2 minutes may be too small for a person to perform the task. Thus, more than one server will be required to reduce the effective f/d time to an allowable value. Management judgment would have to prevail at this point, but we proceed by assuming a server could accomplish the fixturing task in about 6 minutes (.1 hour), and the defixturing task in about 3 minutes (.05 hour), and so estimate for the f/d station

Number of servers = $(18 \times .1) + (18 \times .05) =$
2.7, rounded up to 3

Finally, an estimate of transport time is needed to use the n-o-q models. At least two transports are required per setup, or at least $(6 \times 3 \times 2)$, or 36, transports/hour. In addition, each of the first two setups requires one intermachine transfer, for a minimum of $(36 + 6 + 6)$, or 48, transports/hour. Thus, the maximum transfer time is 1/48, or .021, hour (1.25 minutes). If the transport speed is roughly 50 ft/minute, and the machines are an average of 100 ft from the f/d area, then the average transfer takes about 2 minutes. Again, an increase in the number of servers could be

used to reduce effective transport time, and so let us estimate
that

$$48 \times \frac{1}{30} = 1.6, \text{ rounded up to 2}$$

transporters are needed.

Since the above calculations treat the f/d area and the transportation system as separate stations, the resulting estimates
are also applicable to assumption 2 and its "pipeline" interpretation. (Though, for example, there would be four stations for
f/d in the pipeline, the total of the required number of servers
over the four stations would still be 2.7, as above.)

10.2.3 n-o-q Model Results from Rough-Cut Estimates

The rough-cut estimates of FMS size under either of the above
assumptions can now be refined by analysis using n-o-q models.
The data obtained to this point can easily be placed into the
proper format for, say, the CAN-Q model. Let us do so for
each of the two interpretations for fixtured parts previously
considered.

The CAN-Q data file for the assumption 1 sizing is shown in
Table 10.1. The three setups are represented by three stations
for performing machining, and in addition there is included one
station for f/d and one for transport. The number of servers
and service times for each station are as already calculated. The
operations for the two part types are shown in terms of the three
setups and associated f/d operations (note that the f/d operations are specified as being performed three times for each part's
route through the FMS). Only the number of parts to circulate
in the system has not been specified by the rough-cut method.
Here, we use Solberg's suggestion [1] that the number of parts
in the system be about three times the number of machines.

The output for this initial data set is shown in Table 10.2.
Projected system performance is very close to the desired value
of 6 parts/hour. Station utilizations are relatively high, with
the bottleneck station being the one that performs setup 2. This
is significant since the calculations for the f/d station and for
the transport station were more tentative, and yet they seem to
be sized appropriately. One may argue that the utilization of the
station for setup 1 is a little low (the average number of busy
servers there is less than two) and so one server could possibly

Table 10.1 CAN-Q Input Data for FMS Sizing
(Assumption 1) (Two-Type Example)

1110		
4	30	2
cart	2	2.
fix	3	
set 1	3	
set 2	4	
set 3	4	
typ 1	5	.5
1	6.0	3.
2	21.0	
3	34.8	
4	27.6	
1	3.	3.
typ 2	5	.5
1	6.	3.
2	19.8	
3	40.8	
4	41.4	
1	3.	3.
#eor		

be removed. The output for a second run of the model with this modification is shown in Table 10.3. Indeed, the station utilizations are all high for this design, but note that the system performance has suffered somewhat. Tradeoffs of this type can only be resolved via the intuition of the designer.

The "pipeline" approach to system sizing yields the CAN-Q data file seen in Table 10.4. Though the details of this data set are different from the one for assumption 1, the same general comments apply. The output for this model run is shown in Table 10.5. The production rate is a little lower than desired, but more important from our perspective, the bottleneck station is the transport station. If we increase the number of servers there by one, we find better utilizations for the stations and a much improved production rate (to over 6 parts/hour). However, the bottleneck station is now the f/d area. Trying to increase the number of servers there by one seems reasonable, and the result is shown in Table 10.6. Again, the production rate increases to its highest level so far, and the station utilizations are relatively high.

Table 10.2 CAN-Q Output Data for FMS Sizing
(Assumption 1) (Two-Type Example)

System performance measures

Production rate = 5.829 items/hour

Production rates by product type

	Number	Value
typ1	2.914	2.914
typ2	2.914	2.914

Total value =   5.829

Average time in system = 308.82 minutes

Processing	119.70
Traveling	18.00
Waiting	171.12

Station performance measures

Station number	Station name	Server utilization	Average no. of busy servers
1	fix	0.874	2.623
2	set1	0.661	1.982
3	set2	0.918	3.672
4	set3	0.838	3.351
5	cart	0.874	1.749

10.3 SENSITIVITY OF PERFORMANCE MEASURES FOR n-o-q MODELS

Configuring the FMS as in the previous section is an early activity in FMS design. It is not a comprehensive view of the configuration since some important factors such as cost have not been considered, and other factors have been considered as fixed when in actuality their values may be somewhat arbitrary. Regarding

Table 10.3 Second Run of CAN-Q For Assumption 1
Sizing (Two-Type Example)

System performance measures

Production rate = 5.497 items/hour

Production rates by product type

	Number	Value
typ1	2.749	2.749
typ2	2.749	2.749
Total value =		5.497

Average time in system = 327.44 minutes

Processing	119.70
Traveling	18.00
Waiting	189.74

Station performance measures

Station number	Station name	Server utilization	Average no. of busy servers
1	fix	0.825	2.474
2	set1	0.935	1.869
3	set2	0.866	3.463
4	set3	0.790	3.161
5	cart	0.825	1.649

the latter, the machining times for the part types in the example
FMS were specified as fixed values. Would system performance be
aided significantly by reducing one or more of these times? Perhaps this could be accomplished by using different tooling. Or
would the use of cheaper tooling and longer machining times reduce performance too much to justify the savings? The sensitivity
measures derived in this section can give a designer guidance with
respect to these increasingly more detailed design considerations.

Table 10.4 CAN-Q Input Data for "Pipeline"
Sizing Approach (Two-Type Example)

1110		
6	30	2
cart	3	2.
fix	4	
op 10	2	
op 20	1	
3040	4	
op 50	1	
6070	4	
typ 1	7	.5
1	6.0	3.
2	17.4	
3	3.6	
4	31.8	
5	3.0	
6	27.6	
1	3.	3.
typ 2	7	.5
1	6.	3.
2	16.2	
3	3.6	
4	37.2	
5	3.6	
6	41.4	
1	3.	3.
#eor		

Sensitivity information is presumably used by a system designer to assess the desirability of making changes in controllable parameters of the system. For example, a given system
alternative may display insufficient throughput rate, and as will
be shown below, calculations may indicate that the throughput
is quite sensitive to one or more of the system parameters. These
sensitive parameters then represent candidate changes that the
designer could consider. Note, however, that the emphasis here
is on the controllability of the parameters. If throughput is sensitive to a parameter that the designer has no ability or author-
ization to change, then the sensitivity of that parameter is rather

Table 10.5 CAN-Q Output Data for "Pipeline"
Sizing Approach (Two-Type Example)

System performance measures

Production rate = 5.328 items/hour

Production rates by product type

	Number	Value
typ1	2.664	2.664
typ2	2.664	2.664

Total value =   5.328

Average time in system = 337.81 minutes

Processing	119.70
Traveling	22.00
Waiting	196.11

Station performance measures

Station number	Station name	Server utilization	Average no. of busy servers
1	fix	0.799	2.398
2	op10	0.746	1.492
3	op20	0.320	0.320
4	3040	0.766	3.064
5	op50	0.293	0.293
6	6070	0.766	3.064
7	cart	0.977	1.954

---

useless information. So, in viewing the sensitivity formulas given below, it is important to notice both the system performance measures that are subject to sensitivity study and the parameters whose variation is calculated to change the performance measures.

Table 10.6 Final Run of CAN-Q for Pipeline
Approach (Two-Type Example)

System performance measures

Production rate = 6.239 items/hour

Production rates by product type

	Number	Value
typ1	3.119	3.119
typ2	3.119	3.119
Total value =		6.239

Average time in system = 288.53 minutes

Processing	119.70
Traveling	22.00
Waiting	146.83

Station performance measures

Station number	Station name	Server utilization	Average no. of busy servers
1	fix	0.702	2.807
2	op10	0.873	1.747
3	op20	0.374	0.374
4	3040	0.897	3.587
5	op50	0.343	0.343
6	6070	0.897	3.587
7	cart	0.762	2.287

In the previous chapter, we introduced the classical approach
to n-o-q models via the derivations of Williams and Bhandiwad
[2]. They have also gone on to derive sensitivity formulas for
the system performance measures. Our development here will
be focused on the sensitivity of performance measures with regard to one system parameter, namely the service time at FMS
stations.

Table 10.6 (continued)

Summary for station number 2 : op10

No. of servers	Server utilization	Average no. of busy servers
2	0.873	1.747

Steady state average number of

Items at station	5.745
Items in process	1.747
Items waiting	3.998

Average time spent at this station	Per operation	Per item
Total time (minutes)	55.248	55.248
Processing	16.800	16.800
Waiting	38.448	38.448

Fraction of time	x Items at station	x Items exceeded
x = 0	.0662	.9338
x = 1	.1207	.8131
x = 2	.1095	.7035
x = 3	.0989	.6046
x = 4	.0889	.5157
x = 5	.0795	.4362
x = 6	.0706	.3655
x = 7	.0623	.3032
x = 8	.0546	.2487
x = 9	.0474	.2013
x = 10	.0407	.1606
x = 11	.0345	.1261
x = 12	.0290	.0971
x = 13	.0239	.0732
x = 14	.0194	.0539
x = 15	.0154	.0385
x = 16	.0119	.0266
x = 17	.0089	.0177
x = 18	.0065	.0112
x = 19	.0045	.0067
x = 20	.0030	.0038
x = 21	.0018	.0020

Recall from the previous chapter Buzen's equation for the calculation of the $G(i,j)$:

$$G(i,j) = G(i - 1,j) + X_i G(i,j - 1)$$

This recursive relation was derived using factors of a polynomial $g(t)$, where each factor represented one station, say k. Clearly, the expansion of $g(t)$ is unaffected by rearrangement of $g(t)$'s factors. Let station k's factor be the last (rightmost) one. Then,

$$G(M,j) = G(M - 1,j) + X_k G(M,j - 1)$$

Note that $G(M - 1,j)$ contains no terms involving X_k since it is the product of all factors up to that for the X_k variable. So

$$\frac{\text{del } G(M,j)}{\text{del } X_k} = G(M,j - 1) + X_k \frac{\text{del } G(M,j - 1)}{\text{del } X_k}$$

Since now all G terms are for M stations, we can drop showing the dependence on M. The above equation has the solution

$$\frac{\text{del } G(j)}{\text{del } X_k} = \frac{Q_k(j) \, G(j)}{X_k}$$

In the previous chapter, we showed that the utilization for station i is given as

$$U_i(N) = \frac{X_i \, G(N - 1)}{G(N)}$$

Then, it follows that the utilization at station i varies with a change in relative utilization at station k as follows:

$$\frac{X_k}{U_i(N)} \frac{\text{del } U_i(N)}{\text{del } X_k} = -\left[Q_k(N) - Q_k(N - 1) \right] \quad i \neq k$$

But we would like to see the sensitivity of station utilization in terms of more directly controllable parameters of the system. Recall that

$$X_k = \rho_k \, \tau_k$$

Using the chain rule of derivatives, we can derive

$$\frac{del \ U_i}{del \ \tau_k} = \frac{del \ U_i}{del \ X_k} \frac{del \ X_k}{del \ \tau_k} = - \frac{U_i(N)}{X_k} \left[Q_k(N) - Q_k(N-1) \right] \rho_k$$

$$= - \frac{U_i(N)}{\tau_k} \left[Q_k(N) - Q_k(N-1) \right]$$

Finally, we have an analytical expression for the variation of utilization of station i with change in processing time at station k. Note that this sensitivity depends on the mean number at station i for N and (N - 1) parts in the system. We assume that queue lengths increase proportionately with an increase in number in the system, for closed systems. Therefore, the above derivative must be either negative or zero.

Similar expressions are derived in Williams and Bhandiwad for the sensitivity of other performance measures with regard to processing time.

Consider the slightly different notation used in the CAN-Q methodology. There the production rate P of an FMS was calculated as

$$P = q_M \, w_M \, \frac{G(N-1)}{G(N)}$$

Then, the rate of change of system production rate with respect to processing time at station i can be derived similarly to the above as

$$\frac{del \ P}{del \ \tau_i} = - \frac{P}{\tau_i} \left[Q_i(N) - Q_i(N-1) \right]$$

In fact, this sensitivity of system production rate with regard to station processing time is part of the standard output of the CAN-Q package.

In an analogous manner to the classical approach, we could develop sensitivity measures for the OpA approach. Some of these formulas are given in Suri [3]. This, of course, allows

the extension of OpA methods to a prescriptive environment in
which to aid an FMS designer.

10.4 INTERPRETATION OF SENSITIVITY
RESULTS

The parameters of the n-o-q models are essentially of two basic
types: branching parameters, representing visits by parts to
stations, and service parameters, representing service times and
number of servers at stations. The above formulas provide sensitivity with respect to service parameters, and in particular with
regard to service time sensitivity. Practically speaking, varying
service times in an n-o-q model may relate to a number of changes
in the real-world system. Since classical n-o-q models are more
tractable for the single-part-type situation, recall that a station
service time in the model is actually "averaged" over multiple part
classes if they exist. For the single-part-type case, a service
time change at a station may represent a change in the time of
one or more operations for the part type on that station. These
operation time changes may be further interpreted as changes in
the actual cutting times, or changes in the setup time, or changes
in the tool placement time. For the multiple-part-type case, a
service time change has a more ambiguous interpretation. Since
part type information is being aggregated, a change in service
time may relate to a combination of a change in actual cutting
time for one part type and a change in tool placement time, say,
for another part type. Furthermore, recall that the aggregate
service time in this case also depends on visit frequency of each
part type. Thus, a change in service time could even relate to
a change in the mix of parts in the system.

10.5 SENSITIVITY DATA TO AID FMS DESIGN

Earlier, we showed a rough-cut method to begin configuring the
FMS. That method resulted in an estimate of the number of servers at each FMS station, but did so without estimating the influence of cost on the design. That is, the issue of whether the
number of servers calculated to give desired performance was indeed cost-effective was ignored. In this section, work is described that deals with this issue.
    The previous section considered sensitivity with respect to
service times in an FMS. The sensitivity of FMS performance

measures can also be derived with respect to number of servers
at each station. This is a more difficult derivation, but the parameter in this case is also more significant to preliminary design
phase studies. At this stage of design, even the required number of machines in the FMS may not be obvious. A sensitivity
measure regarding number of servers could provide useful information for such a decision. Such a measure has been derived by
Solberg and used by Yu [4] to investigate the cost-performance
tradeoff as the number of machines in an FMS is varied.

The focal points of Yu's study were the production rate of an
FMS design and its associated cost. In particular, the sensitivity of the production rate, P, with respect to the number of servers at each station i, denoted $pdel_i$, served as the vehicle to
improved design. $Pdel_i$ is calculated from n-o-q model (i.e.,
CAN-Q) results in one pass by means of an undisclosed equation. More precisely, $pdel_i$ is defined as the change in production rate with an increase (alternatively, decrease) in the number of servers at station i by one. As usual, the calculation is
made for a closed system and is exact only for the value of N
parts in the system for which the program was run. It is interesting to note that, among the various stations, the one with the
highest value of pdel is not necessarily the one that showed highest utilization.

Given a pdel-augmented CAN-Q program, it was relatively
straightforward to incorporate it into an iteration loop to search
efficiently for more productive system configurations. Here, a
"configuration" refers only to the number of machines at each
station in the FMS. More detailed considerations that may be
associated with the term "configuration," such as physical location of machines and distances between them, are not represented
in n-o-q models.

The iteration process in this study begins with a minimum configuration, say one machine at each station, where the number of
stations has been prespecified (we have already shown examples
of how this number of stations can be estimated). This configuration is submitted to a CAN-Q run obtaining the production
rate and the sensitivity values, $pdel_i$. A reasonable approach
to production rate improvement is then to increase the number
of servers at the station that has the most positive pdel value.
However, an increased number of servers implies that more
parts can be served by the system. Thus, the iteration program
also adjusts the number, N, of parts in the system for the new
number of servers. N is, in fact, set so as to achieve 95% of
the maximum production rate for that configuration. From this

Table 10.7 Optimal FMS Configurations via Modified CAN-Q

Index	P	C	T	U	N	Next	Pdel⁺	NS				
1	1.91	14.00	377.06	.68	12	3	1.91	1	1	1	1	1
2	2.30	16.00	312.68	.65	12	4	.39	1	1	2	1	1
3	2.85	17.00	252.61	.67	12	2	.55	1	1	2	2	2
4	3.81	21.00	283.58	.77	18	3	.96	1	2	2	2	2
5	4.57	23.00	367.53	.81	28	4	.76	1	2	3	2	2
6	4.77	24.00	289.05	.75	23	5	.20	1	2	3	3	3
7	5.57	26.00	366.30	.88	34	1	.79	1	2	3	3	3
8	5.71	31.00	315.41	.81	30	2	.14	2	2	3	3	3
9	5.71	35.00	178.66	.74	17	3	.00	2	3	3	3	3
10	6.88	37.00	209.40	.81	24	4	1.17	2	3	4	3	3
11	7.61	38.00	204.98	.83	26	3	.73	2	3	4	4	4
12	8.55	40.00	245.59	.87	35	2	.94	2	3	5	4	4
13	9.14	44.00	262.63	.86	40	4	.59	2	4	5	4	4
14	9.50	45.00	252.62	.84	40	5	.36	2	4	5	5	5
15	9.54	47.00	176.07	.84	28	3	.04	2	4	5	5	5
16	11.31	49.00	323.70	.94	61	2	1.76	2	4	6	5	5
17	11.41	53.00	289.20	.90	55	1	.10	2	5	6	5	5
18	11.42	58.00	204.96	.85	39	4	.01	3	5	6	5	5

Index, index number for alternative; P, production rate; C, system cost; T, average time in system; U, average utilization; N, number of parts in system; Next, station number with max pdel⁺; Pdel⁺, magnitude of maximum production increase; NS, number of servers at each station.

point, iteration continues until a user-specified maximum production rate or system cost is achieved.

The Yu iteration process provides a list of configurations with
associated production rates and costs, as in Table 10.7. The
system designer can select a configuration that meets production or cost constraints, depending on which aspect of performance has higher priority. Other aspects of performance may also
be included in the selection decision. For instance, note that for
configurations indexed 8 and 9, the production rate is identical,
but the cost is higher for 9. However, it is also the case that
the number of parts in the system (i.e., in-process inventory)
is considerably lower for 9. Selection of a system configuration
according to these multiple, and conflicting, criteria is best left
to the intuition of the designer.

Yu's study is interesting in another regard. If an exhaustive
enumeration of configurations is made, a graph of production rate
versus cost is obtained as in Fig. 10.1. Note the "plateaus" in

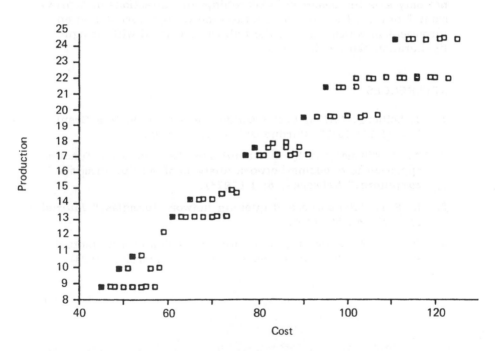

Fig. 10.1 Production rate versus cost for FMS configurations.

the graph where extra stations are added (and so extra costs incurred) without much increase in production rate. The left-hand
point at each plateau represents a dominant configuration with
respect to the others on the same plateau. Yu's iteration scheme
apparently finds these dominant points. She also reports that
such points are rather insensitive to the form of the cost models,
provided only that they are monotonically increasing with the number of stations in the system.

10.6 SUMMARY

The early stages of FMS design are carried out in a context of ill-
defined system specifications. Design tools are needed that require the minimal amount of data in order to accomplish at least
the most essential tasks of design, such as system sizing. The
n-o-q models serve very well in this capacity.

The n-o-q models have the additional capability of providing
sensitivity measures for certain of the system parameters. This
not only aids the designer in searching for "directions of improvement," but can lead to a more automated design workstation environment in which much of the tedium associated with design-
by-iteration can be alleviated.

REFERENCES

1.   J. Solberg, *CAN-Q User's Guide*, report no. 9, NSF Grant
     No. APR 74 15256, Purdue University (1980).

2.   A. C. Williams and R. A. Bhandiwad, "A generating function
     approach to queueing network analysis of multiprogrammed
     computers," *Networks, 6*: 1 (1976).

3.   R. Suri, "Robustness of queuing network formulas," *Journal
     of ACM, 30*: 564 (1983).

4.   D. Yu, "A flexible technique for the design of manufacturing
     systems," M.S. degree thesis, Purdue University (1983).

11

Simulation for FMS Design

11.1 SIMULATION MODELS BASED ON
n-o-q MODELS OF FMS

The previous chapter presented an approach to FMS design that can aid the design process through its preliminary stages. As the designer contemplates more detailed considerations, he must turn to a type of model that can accommodate that increased level of detail. Simulation is such a robust descriptive tool. To illustrate the further specification of an FMS design that may require simulation modeling, take the example two-part-type system subject to preliminary design by rough-cut and then n-o-q methods in the previous chapter (in particular, consider the design that resulted from the "pipeline" or assumption 2 approach).

To give direction to our progression from n-o-q models to simulation models, consider the data-driven type of simulation procedure. These are closest to n-o-q models in the sense that both require only numerical data. Our discussion will be further facilitated if we focus on the GCMS simulator and its input forms, presented in Chapter 8. Proceeding through those forms one statement type at a time, we can systematically learn about the need for increased detail in system specification in order to use simulation.

As noted previously, the GCMS input data (or "model") for this example consists of rows of numbers, each row representing numerical data for the part types to be produced (including process plan and operation times), or for the fixture or pallet types available for the parts, or for the workstations in the FMS, or for the material-handling-system characteristics. The format for each type of data row is given in Table 8.2.

The GCMS input data are formatted, as was CAN-Q input, by a sequence of groups of data statements. Each group consists of one or more statements, and the groups must be listed in a given order. That order is shown in Table 8.2. An examination of those tables indicates that simulation modeling can represent an FMS's behavior in at least the following additional aspects that were ignored or assumed away in earlier n-o-q analysis:

1. The physical layout of the system, including showing multiple machines of each type as separate entities, specifying the distances between machines, and allocating fixtures among the separate fixture/defixture stations
2. Explicitly representing the material-handling (mh) devices with regard to their capacity to hold parts, their speed of movement, and the routes they may take in their travel
3. The finite capacity of waiting areas at the machines, as well as the type of shuttle mechanism used to move parts into and out of those areas
4. Various decision rules for selection of parts to be processed and for machines to process them on, as well as for choice of mh device to transport them between stations

The GCMS SYSTEM statement is the first statement in the data set and immediately requires the designer to expand the FMS's specification by requesting information on the number of workstations in the system (see field 6 of this statement type in Table 8.2). The n-o-q models aggregated separate machines into single stations, both for machining and for fixture/defixture operations. Now these data must be disaggregated by machine.

The mh system must be described in more detail. In particular, its magnitude with regard to number of decision points (places where mh devices may be stopped or rerouted) and with respect to number of mh devices must be stated. This implies the existence of a layout for the FMS, including the placement of all machines, the physical distances between them, and the paths that mh devices may take when transporting parts. In fact, that layout information is used in the GCMS WORKSTATION, MH DEVICE, and DECISION POINT statements to be considered below.

The OPERATION statements of GCMS seem similar to those for n-o-q models. However, it should be noted that if several stations perform a given operation, one station can be given priority over others. In addition, on the PART TYPE statement, a higher priority can be assigned to one part type over another, where this assignment can affect the way parts are selected from queues for processing (see field 7 of the WORKSTATION statement in Table 8.2).

WORKSTATION statements also demand new data from the designer. A selection of the size of available waiting spaces must be made for both arriving and departing parts at each machine. Moreover, the location in the FMS layout for each of these areas must be specified in terms of the mh decision points. If the machines are subject to failure, the statistics associated with that unreliability must be specified.

Finally, on many of the statements, there is a request for choice of a decision rule which represents some of the operating policy of the FMS. These decision rules include (simple) procedures for finding the next operation for a part, for finding the next station to perform the operation, for finding an mh device to move the part, and for selecting parts from a queue of those that have arrived to be processed.

With the above thoughts on additional information in mind, study the GCMS input data for the associated two-part-type FMS, as given in Table 11.1. One possible layout for the FMS which is consistent with that information is shown in Fig. 11.1. As a result of the need to disaggregate the data so that individual machines can be described, the number of workstations specified to GCMS is 15 (12 machines and three fixture/defixture stands). Note that the number of operations has been expanded to 22 (from 14 for the n-o-q model). This has resulted from disaggregation of the two operations labeled 30 and 40, as well as those labeled 60 and 70. Furthermore, a separate operation is now specified for each fixturing. The fixturing operations have been (arbitrarily) assigned to separate fixture/defixture stations. The machining operations have also been assigned to multiple machines, implying that those machines are identically tooled. For example, operation 10 is to be done on either machine 1 or 2. The decision as to which machine is selected is specified in field 3 of that operation's statement, where the 2 indicates that the station will be selected on the basis of which machine is currently idle. If both are busy, then the part will be sent to machine 1 (it is listed first).

The PART TYPE data indicate that, at most, 15 parts of each type may be in the system at any one time. This limit (and the total of 30) is consistent with the number of parts allowed in the

Table 11.1 GCMS Input Data for Two-Part-Type FMS

```
two part type fms,22,2,6,30,15,1,25*
pt1 fx1,1,1,13,6.*
pt1 op 10,2,2,1,17.4,2,17.4*
pt1 op 20,3,1,3,3.6*
pt1 fx2,4,1,14,9.*
pt1 op30,5,2,4,9.,5,9.,6,9.,7,9.*
pt1 op 40,6,2,4,22.8,5,22.8,6,22.8,7,22.8*
pt1 op 50,7,1,8,3.0*
pt1 fx3,8,1,15,9.*
pt1 op 60,9,2,9,14.4,10,14.4,11,14.4,12,14.4*
pt1 op 70,10,2,9,13.2,10,13.2,11,13.2,12,13.2*
pt1 fin, 11,1,13,3.*
pt2 fx1,12,1,13,6.*
pt2 op 10,13,2,1,16.2,2,16.2*
pt2 op 20,14,1,3,3.6*
pt2 fx2,15,1,14,9.*
pt2 op 30,16,2,4,9.,5,9.,6,9.,7,9.*
pt2 op 40,17,2,4,28.2,5,28.2,6,28.2,7,28.2*
pt2 op 50,18,1,8,3.6*
pt2 fx3,19,1,15,9.*
pt2 op 60,20,2,9,19.8,10,19.8,11,19.8,12,19.8*
pt2 op 70,21,2,9,21.6,10,21.6,11,21.6,12,21.6*
pt2 fin,22,1,13,3.*
part ty 1,1,1,15,1,1,1,48,11,1,2,3,4,5,6,7,8,9,10,11*
part ty 2,2,2,15,1,1,1,48,11,12,13,14,15,16,17,18,19,20,21,22*
pal ty 1,1,15,13,1*
pal ty 2,2,4,14,1*
pal ty 3,3,4,15,1*
pal ty 4,4,15,13,2*
pal ty 5,5,4,14,2*

dec pt 13,13,7,0,12,10.,,14,30.*
dec pt 14,14,8,0,13,30.,15,30.*
dec pt 15,15,9,0,14,30.,16,10.*
dec pt 16,16,9,0,15,10.,17,30.*
dec pt 17,17,10,0,16,30.,18,10.*
dec pt 18,18,10,0,17,10.,19,30.*
dec pt 19,19,11,0,18,30.,20,10.*
dec pt 20,20,11,0,19,10.,21,30.*
dec pt 21,21,12,0,20,30.,22,10.*
dec pt 22,22,12,0,21,10.,23,30.*
dec pt 23,23,13,0,22,30.,24,20.*
dec pt 24,24,14,0,23,20.,25,20.*
dec pt 25,25,15,0,24,20.,1,30.*
mh 1,1,1*
mh 2,2,1*
mh 3,3,1*
mh 4,4,1*
mh 5,5,1*
mh 6,6,1*
3 8-hour shifts, 1440,400,450*
```

```
pal ty 6,6,4,15,2*
stat 1,1,2,2,2,0,1,.1,.1,1,2*
stat 1,2,2,2,2,0,1,.1,.1,3,4*
stat 2,3,2,0,4,1,.1,.1,5,5*
stat 3,4,4,2,2,0,1,.1,.1,6,7*
stat 3,5,4,2,2,0,1,.1,.1,8,9*
stat 3,6,4,2,2,0,1,.1,.1,10,11*
stat 3,7,4,2,2,0,1,.1,.1,12,13*
stat 4,8,2,0,4,1,.1,.1,14,14*
stat 5,9,4,2,2,0,1,.1,.1,15,16*
stat 5,10,4,2,2,0,1,.1,.1,17,18*
stat 5,11,4,2,2,0,1,.1,.1,19,20*
stat 5,12,4,2,2,2,0,1,.1,.1,21,22*
lu 1,13,4,0,6,1,.1,.1,23,23*
lu 2,14,2,0,6,1,.1,.1,24,24*
lu 3,15,2,0,6,1,.1,.1,25,25*
mh ty 1,1,6,1,50,1,1,2,3,4,5,6,7,8,9,10,11,12,13,14,15,16,17,18,19,20,21,22,23
24,25*
dec pt 1,1,1,0,25,30.,2,10.*
dec pt 2,2,1,0,1,10.,3,30.*
dec pt 3,3,2,0,2,30.,4,10.*
dec pt 4,4,2,0,3,10.,5,30.*
dec pt 5,5,3,0,4,30.,6,30.*
dec pt 6,6,4,0,5,30.,7,10.*
dec pt 7,7,4,0,6,10.,8,30.*
dec pt 8,8,5,0,7,30.,9,10.*
dec pt 9,9,5,0,8,10.,10,30.*
dec pt 10,10,6,0,9,30.,11,10.*
dec pt 11,11,6,0,10,10.,12,30.*
dec pt 12,12,7,0,11,30.,13,10.*
```

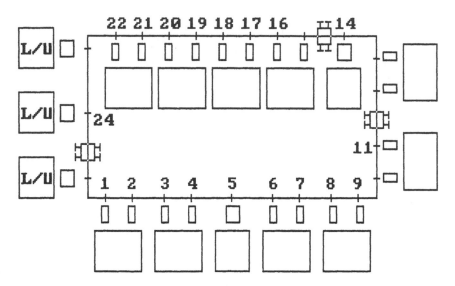

Fig. 11.1 Layout for two-part-type FMS.

associated closed-network n-o-q model given in the previous chapter. The production target for each part type specifies only the ratio of part mix of completed parts, so that equal numbers in field 8 of each part type indicate that the part mix is to be split evenly.

The size of the waiting areas at the machines, required on the WORKSTATION statements (fields 4–6 in Table 8.2), was initially estimated from the queue size data out of the n-o-q analysis. For example, for the station labeled "stat 1," which performs operation 10, the n-o-q data for queue length distribution showed that the waiting areas had four or more parts in them about 50% of the time (see the "fraction of time waiting" data in Table 10.6). This was converted over to the simulation data by dividing that queue size by the number of stations, in this case $4/2 = 2$, and then rounding up. This gave what turned out to be reasonable estimates for all except the fixture/defixture stands. There, the n-o-q estimates (of size 1) were far too low because those stands were specified in the n-o-q model in terms of parallel servers. The final queue sizes of all stations were determined by trial-and-error such that a simulation run of three shifts' (24 simulated hours) duration did not experience a deadlocked system. A deadlocked system results, for example, when all waiting positions at a machine are filled, a part is done on the machine (but cannot be moved off), and the mh devices are all waiting to drop off parts at that station.

Most of the FMS layout information is contained in the DECISION POINT statements. Physical distances are stated here, in addition to the presence of machines at each of the decision points. Note that the mh system is specified as having two-way paths by mentioning both the points adjacent to any point as being successors.

Running the GCMS model for this FMS for two shifts (not started idle and empty, since data for an initial shift was discarded), gives the output shown in Table 11.2. Even with all queues sized so that no deadlocks occur, the production rate for the simulated system is only about 5 parts/hour, as compared to our desired rate of 6 parts/hour. Perhaps a little more "tuning" of the system parameters could bring the production rate closer to the desired level. Some possibilities are to revise the fixture/defixture area to provide truly parallel service, to increase the number and/or speed of mh devices, or even to modify the configuration of the system by adding new "cross-branches" to the existing mh device paths.

11.2 UNCONTROLLABLE FACTORS IN DESIGN

The increased complexity of FMS entails an enormous number of design alternatives, each of which will meet a designer's objectives to a greater or lesser degree. Each such alternative corresponds to the designer's selection of a set of system "parameters," e.g., number of machines, type of machines, number of tools, placement of tools, number and type of mh devices, and so forth. Presumably each parameter has some effect on the FMS performance measure values (otherwise its choice is arbitrary). Since the designer chooses these parameters, we may say that they are controllable factors of the design. But there are other factors which affect FMS performance and which are largely beyond the designer's control. For example, machines in the system may break down from time to time. Generally, such breakdowns will significantly affect system performance. Yet the designer cannot specify when such events will occur (obviously,

Table 11.2 GCMS Output for Two-Part Type FMS

Production summary for completed parts

Part type	Production		For parts completed time in system		
	Parts compl	%	Ave	Min	Max
1	40	62.5	277.19	205.70	471.40
2	38	66.7	276.62	198.10	414.70
Total	78				

Station performance summary

Station no.	Time busy	%	Time idle	%	Time down	%	% of time busy during time available
1	1290.20	89.6	149.80	10.4	0.00	0.0	89.60
2	504.00	35.0	936.00	65.0	0.00	0.0	35.00
3	297.70	20.7	1142.30	79.3	0.00	0.0	20.67
4	1405.90	97.6	34.10	2.4	0.00	0.0	97.63
5	1153.20	80.1	286.80	19.9	0.00	0.0	80.08
6	1008.40	70.0	431.60	30.0	0.00	0.0	70.03
7	586.20	40.7	853.80	59.3	0.00	0.0	40.71
8	262.20	18.2	1177.80	81.8	0.00	0.0	18.21
9	1354.20	94.0	85.80	6.0	0.00	0.0	94.04
10	1141.20	79.3	298.80	20.7	0.00	0.0	79.25

11	1088.40	75.6	351.60	24.4	0.00	0.0	75.58	0
12	1039.60	72.2	400.40	27.8	0.00	0.0	72.19	0
13	1177.30	81.8	262.70	18.2	0.00	0.0	81.76	4
14	1217.90	84.6	222.10	15.4	0.00	0.0	84.58	0
15	719.60	50.0	720.40	50.0	0.00	0.0	49.97	0

Shuttle performance summary

Station no.	On-shuttle			Off-shuttle			Generalized		
	Queue size	Ave que	Max que	Queue size	Ave que	Max que	Queue size	Ave que	Max que
1	2	0.836	2	2	0.189	2	0	0.000	0
2	2	0.006	1	2	0.124	2	0	0.000	0
3	0	0.000	0	0	0.000	0	4	0.754	4
4	2	1.336	2	2	0.342	2	0	0.000	0
5	2	0.309	2	2	0.239	2	0	0.000	0
6	2	0.204	2	2	0.247	2	0	0.000	0
7	2	0.158	2	2	0.151	1	0	0.000	0
8	0	0.000	0	0	0.000	0	4	0.508	3
9	2	0.799	2	2	0.236	2	0	0.000	0
10	2	0.217	2	2	0.172	1	0	0.000	0
11	2	0.177	2	2	0.194	2	0	0.000	0
12	2	0.123	2	2	0.206	2	6	0.528	4
13	0	0.000	0	0	0.000	0	6	0.987	6
14	0	0.000	0	0	0.000	0	6	0.547	3
15	0	0.000	0	0	0.000	0			

(continued)

Table 11.2 (Continued)

Cart performance summary

Move time	%	Waiting, time	%	Down time	%	Idle time	%	Total distance moved	Number of assignments
				Cart activities					
1308.50	90.9	32.19	2.2	0.00	0.0	99.3	6.9	36510.00	159
1265.40	87.9	29.79	2.1	0.00	0.0	144.8	10.1	35130.00	147
1275.80	88.6	27.59	1.9	0.00	0.0	136.6	9.5	30110.00	135
1248.00	86.7	27.90	1.9	0.00	0.0	164.1	11.4	30960.00	137
724.30	50.3	116.40	8.1	0.00	0.0	599.3	41.6	21210.00	88
603.70	41.9	113.50	7.9	0.00	0.0	722.8	50.2	24410.00	107

if he could, he would specify them to never occur). So effective design of an FMS results in one or more alternative designs which allow the FMS to meet desired performance objectives in the face of the presence of (some) uncontrollable factors. We shall focus on three of the more significant uncontrollable factors called, for convenience, u-factors.

11.3 TYPES OF U-FACTORS IN FMS

Experience with both simulating and actually operating FMS indicates that at least three factors are beyond the close specification of the designer, yet they do affect the performance of the FMS. These u-factors include human operation times, arrival and introduction patterns for the raw workpieces to the FMS, and breakdowns of machines, tools, and other components within the FMS.

The operation times for humans involved in the system (e.g., in fixturing operations) cannot be dictated by the designer, though they may vary significantly from human to human, or from time to time for one operator. The second type of u-factor, flow of work into the FMS, is combinatorial in nature. The arrival of orders is a multidimensional phenomenon. Incoming orders could be for parts of various types, for various quantities, with various due dates, and, to add complication, are usually produced in a mix simultaneously with other part types (this, in fact, is the rationale for FMS!). Thus, any thorough investigation of the effects of this u-factor would need to vary all the dimensions associated with each order and would also need to vary in a realistic way the mix of those orders. Furthermore, the rule for selecting which order to process next would also need to be varied.

An equally complex case is that of system component breakdown. The complexity arises here not only because of the dimensionality of the number of individual and combination breakdowns that can occur, but also as to when they occur and, most important, as to what the FMS control system's response is to the breakdown. The types of breakdowns of components can be enumerated, and even the effects of the timing of the breakdown and subsequent repair can be subject to some systematic study. However, the control strategies for responding to breakdowns are open-ended in that they are limited only by the imagination of the FMS designer.

11.4 CONSIDERATION OF U-FACTORS
IN FMS DESIGN

U-factors can be characterized along many dimensions. If the
designer is to account for their presence, he may find it inefficient to do so by simply enumerating all possible ways in which
the u-factors may occur and then run a simulation for each case.
A more systematic approach to do such accounting could be to
define a set of "typical" occurrences of the u-factors. This
could include two or three anticipated order mixes, say, along
with a couple of typical operation times for a human operator,
all in conjunction with an assumed breakdown pattern for perhaps two machines in the system. The designer could then construct a simulation "experiment" which assesses performance measure values for all possible combinations of this very small set of
possible occurrences. Given the resources to accomplish this experiment, he is still left with the task of deciding which of the
performance values obtained is the "correct" one, if indeed any
one of them is. Or he can average the results. The problem remains that this experiment has considered only a few of the tre-
mendously large number of possibilities for the u-factors (and
there is no guarantee that the "typical" occurrences assumed are
indeed representative of those possibilities).

A second approach is define a range over which each of the
u-factors may vary and then to "randomly" select u-factor values from these ranges. In effect, this approach attempts to
obtain as a result a range of performance measure values, as did
the approach mentioned in the previous paragraph. In this case,
however, the random selection is done according to a set of consistent probability rules rather than via intuition. Therefore,
the resultant set of performance values can also be treated according to the same consistent rules. This second approach is
the one considered in the next sections.

11.5 DESIGN OF EXPERIMENTS ACCOUNTING
FOR U-FACTORS

Generally, one simulation run to describe performance of a simulated system is not sufficient. This is the case because there
are often factors associated with the simulated system that are
subject to considerable uncertainty. If humans work in the system, for example, their time to perform a given operation may

vary considerably over the time period of one shift. Which of the various observed operation times should the simulation program use? All simulation procedures handle this situation by representing the uncertain factor (in this case, the operation time) by use of a submodel. This submodel is usually represented as a sampled statistical distribution (of operation times, say), and the simulation procedures provide various forms of such distributions. By means of the sampling process, it is possible to make two runs of a simulation and obtain different performance estimates from each (i.e., in one run, the human operator times may, on average, be smaller than in another run). So, for the cases where the simulation model contains submodels of uncertainty, the performance of the system obtained from only one run is a random variable itself. One is led to seek averaged performance measure results obtained over possibly many simulation runs. Considerable effort has been devoted to the proper means of achieving these averaged performance measure values. These efforts place the performance analysis task squarely in the domain of experimental design. In this context, the simulation model is seen as an experimental laboratory for which a design of experiments must be made in order to obtain measures in which one can have confidence.

11.6 SIMULATION OUTPUT ANALYSIS UNDER UNCERTAINTY

11.6.1 The Confidence Interval Concept

Statistical performance analysis (or output analysis, according to Law and Kelton [1]) for FMS systems is in fact necessary only in well-defined situations. Since FMS do not usually employ human effort for processing tasks within the FMS, that element of uncertainty is eliminated. Most FMS do, however, require human effort at the fixture/defixture stations since automated fixturing is not yet practiced. If these operation times are pertinent to the system performance measure that establishes the main criterion (e.g., average time to complete an order), then they would need to be represented by statistical submodels. A second element of uncertainty in assessing the performance of FMS is introduced by the arrival of orders and raw workpieces to the FMS. Customer orders, for example, may only be describable by some statistical process that specifies the "usual" part type ordered, the "usual" quantity of product desired,

and "typical" due date. A third important aspect of uncertainty
that significantly affects system performance are the breakdown
events associated with tools, machines, and material-handling
(mh) devices.

Law and Kelton display, for performance analysis, several useful methods which imply that (statistical) confidence in performance assessment under uncertainty can be achieved if one is provided a certain level of computing resources. The methods we
shall consider fall into two categories, the first type to obtain
statistically valid values of a given FMS design's performance,
and the second to properly compare the performance measure
values between two alternative FMS designs.

Imagine that an FMS designer has arrived at a tentative configuration for a proposed system. Some estimate of the system's
production performance is needed to assess the worth of this configuration (as opposed to some other). A first step toward the
performance assessment could be the development of a simulation
model, in any of the forms previously described. If the modeled
system had no elements of uncertainty in it, then one simulation
run would suffice to estimate performance. In reality, however,
each FMS configuration is subject to at least one of the three
types of uncertainty described earlier. So a careful simulation
study would include several simulation runs to obtain an averaged performance estimate.

What is it that changes from one simulation run to the next as
we calculate the FMS performance estimate? Suppose that one of
the uncertainties has to do with the time(s) at which a given machine in the FMS experiences a failure. Such a failure could occur when there are many parts waiting for this machine, or conversely when the parts-waiting area is empty. Obviously, the
former situation adversely affects system performance far more
than the latter one. To "average" these many possible situations,
we make multiple simulation runs. In this case, the thing that is
possibly changing in each run is the time at which the machine
fails. This knowledge entails a further question. What is it about
the mechanics of a simulation model that could represent a machine
failing at one time in one run and failing at a different time in a
second run? The most common means of this representation in
simulation modeling is that of a statistical submodel. That is, the
uncertain event is characterized by a statistical distribution of
some meaningful form. For the case of breakdowns, this distribution usually represents the time interval between failures. But
how are differing time intervals selected? Underlying each simulation run is a sequence (often called a "stream") of numbers,

each number between 0 and 1. This sequence or stream is random in that the value of one number in the sequence is (theoretically) independent of the numbers just adjacent to it (and indeed of all the other values as well). Now, a procedure is provided by the simulation language for converting any one of these
random numbers to, say, the time interval to the next machine
failure. Finally, then, if we change the random-number stream
from one simulation run to the next, we can indeed get a new
pattern of times between failure of the specified machine in the
second run from the first. The simulation languages discussed
in this book all have the facility to modify random-number streams
from one run to the next.

An important consideration to the FMS designer is just how
much the estimated performance might change with a change in
the random-number stream. The expense of a statistical analysis should only be incurred if the performance changes between
runs are significant. To illustrate the possible variation in performance measure values, consider a simple situation where orders (i.e., parts) arrive to some machine and let the order arrival pattern be uncertain. Generally, an uncertain arrival pattern is represented by an exponential distribution. Numerically
controlled machines, however, usually have a specified machining
time with very little uncertainty. Let us assume, then, that orders arrive to the machine according to an exponential distribution with a mean value of 6 minutes between arrivals. Furthermore, say that the machining time for each order is fixed at 5
minutes. If we simulated this situation and changed the random
number stream 10 times, we could collect 10 "observations" of,
for example, the average waiting time for the first 100 completed
orders in each run. One such experiment resulted in the following observations of that average waiting time:

3.87   0.99   1.43   0.84   1.89

0.61   1.88   1.16   1.72   1.24

There is a considerable range of variation in these values. Which
of these values is "correct"? Or is the average of these 10 numbers the "correct" value? Unfortunately, we can never know the
mythical correct value. However, we can find limits within which
we can be fairly certain that this correct value or "true mean"
lies. Those upper and lower limits are obtained rather easily by
computing a confidence interval from the set of observations.

In general, if we have N observations or samples from an experiment, then let us denote the sample mean of the N values by

$X^\frown(N)$. The measure of the range of variation of values obtained from the experiment, called the variance of the sample, is given by

$$s^2[X^\frown(N)] = [\sum_{i=1}^{N} (X_i - X^\frown(N))^2]/(N - 1)$$

Then the confidence interval (L,U) is defined as the interval

$$X^\frown(N) \pm t_{(N-1,\ 1-\alpha/2)} \times \sqrt{[s^2[X^\frown(N)]}$$

where $t_{(N-1,\ 1-\alpha/2)}$ is a number obtained from a table of the values of the t-distribution, with α between 0 and 1. We may interpret this to say that the sentence "the true mean lies between L and U" is a true statement with probability $(1 - \alpha)$. For our experiment, if we set a confidence level of 0.9 (i.e., we want to have 90% confidence), we obtain an interval

$$1.56 \pm 1.83 \times \text{sqrt } \frac{0.844}{10} = (1.03, 2.09)$$

Informally, we can say that if we calculated 10 such confidence intervals (using many different random number streams), then 9 of the 10 should contain the true value for average waiting time of the first 100 parts at the machine.

Consider a second example which is perhaps more meaningful to an FMS designer. Suppose that the two-part-type system as designed earlier is subject to unreliability of one of its machines. In particular, let the machine numbered 4 fail about once every hour and let the repair time for it be about a half-hour (in fact, let the breakdown distribution be normal with mean of 60 minutes and standard deviation of 30 minutes, while the repair is normal-distributed with mean 30 and standard deviation of 15). Parts that would go to machine 4 are diverted to machine 7 when machine 4 is down. All of this reliability information is added to our GCMS model by changing one statement for that model, namely the one for machine 4 to

stat typ 3,4,4,2,3,0,1,.1,.1,6,7,2,60,30,,500,2,30,15,,200,7*

The performance of the two-part-type FMS will, of course, vary depending on the pattern of the breakdowns. To obtain different patterns, we use the fact that GCMS allows six random-number

seeds for its simulation experiments. In this way, we can estimate the system behavior for six different breakdown patterns.

The results from running GCMS are enlightening for two reasons, one being the measurement of uncertainty with regard to
production rate in the presence of breakdowns. The other insight gained, however, has more to do with the effect of breakdowns on the system performance. For example, with an unreliable machine 4, some waiting areas that functioned well for the
nonbreakdown case now fill up and cause system deadlock. So
it was necessary to increase the size of the waiting areas at the
fixture/defixture stations to eight in order to continue our simulation experiments. Similarly, the queue sizes at stations 3
through 7 were also increased to achieve reasonable system operation. In effect, simply running such unreliability experiments
contributed to FMS design.

The GCMS data obtained for total production of parts in three
shifts (starting idle and empty) were

94   94   90   91   98   87

Using the notion of confidence interval to estimate the "true mean"
of these data leads to a 90% confidence interval for parts completed of (89.1, 95.2). Note that even the rather wide range for
the confidence interval does not include two of the data points!
This is an indication of the danger of relying on one computer
run for estimating performance measure values in the presence
of uncertainty.

It may be discomforting for some readers to accept the notion
of a confidence interval with its relatively wide range as opposed
to, say, the mean of the set of observations $X^\frown(N)$. However,
it is important to note that $X^\frown(N)$ can itself vary widely; for example, we could make new runs, if provided with more random
number seeds, and compute their new average value [in fact,
$X^\frown(N)$ is itself a random variable]. Moreover, the width of the
confidence interval can be reduced — by making more simulation
runs and obtaining more observations. The latter suggestion is,
unfortunately, a fact of life associated with proper statistical use
of simulation modeling and experimentation.

11.6.2 Statistical Comparison of Alternative FMS Designs

Above, it is shown that the confidence interval notion is a statistically precise way of estimating the performance measure value
of an FMS when some aspect of the system exhibits uncertainty.

A modification of that notion can be used to compare the difference in performance values of two alternative FMS designs when, again, in each case the performance is subject to uncertainty [1]. Here, let us assume that each alternative design is simulated N times with a different random-number stream each time. This gives N independent observations of performance for each design. Let $X_{1,j}$ denote the observed performance measure value for design 1 on the jth run, and similarly for $X_{2,j}$. Now consider the difference between these observations, $Z_j = X_{1,j} - X_{2,j}$. It may be shown that the Z_j's are independent random variables, and so we can compute a confidence interval for them. That is, let

$$Z^{\sim}(N) = (\sum_{j=1}^{N} Z_j)/N$$

and

$$\text{variance}[Z^{\sim}(N)] = \{\sum_{j=1}^{N} [Z_j - Z^{\sim}(N)]^2\}/N(N-1)$$

and form the (approximate) confidence interval

$$Z^{\sim}(N) \pm t_{(n-1,\ 1-\alpha/2)} \sqrt{(\text{variance } Z^{\sim}(N))}$$

Note that this confidence interval is likely (depending on the value of α) to contain the "true mean" of the Z_j. Then, if this interval does not contain the value zero, there is some confidence to the belief that these two FMS designs are indeed different in their performance. Furthermore, if the confidence interval is entirely positive, there is also confidence that the performance of design 1 is superior to that of design 2.

11.7 AUTOMATED DESIGN VIA SIMULATION

System design using simulation as a tool is essentially an iterative process that takes place in the later stages of design. The designer/modelers use their intuition to arrive at a system configuration that should meet requirements. A model reflecting the configuration is constructed and verified. Simulation runs are made to estimate the performance of the modeled system, and changes in configuration may be suggested by the simulation output. These changes can relate to "parameters" of the system such as speed of mh devices, number of pallets available at load stations, dispatching rules for part movement, picking rules for parts in waiting areas, and so forth. The number of such changes can be enormous, and a systematic search through all these alternatives could be tedious and expensive in terms of the modeler's time and the computer's time. Such a search is further confounded if the designer also wishes to consider the effect of uncontrollable factors. In these cases, the designers must resign themselves to obtaining as satisfactory a choice as possible within the constraints of time and resources available.

There are some tools available that can expedite the designer's search for the best possible system design while using simulation as a design tool. One methodology places the simulation model within a feedback loop with another computer program which plays the role of an automated, and therefore unsophisticated, "computerized designer." This program calls on the simulation in an iterative manner, with new parameters on each iteration, somewhat like that of the human designer. This approach is referred to here as simulation optimization. A second approach obtains sensitivity measures for various system parameters by collecting data from a simulation run in a special way. This approach is known as perturbation methodology. Both approaches can consider the uncertainty associated with uncontrollable factors and are discussed below.

11.7.1 The Performance Function

As indicated in the previous section, some FMS parameters are measured in terms of numbers and so may have an integer nature (e.g., number of pallets) or may be continuous (e.g., speed of mh devices). Other parameters can only be described in a nominal way (e.g., type of picking rule for selecting parts from a queue). Some of the automated design methods we shall consider were initially developed for those cases where the parameters of interest are describable as continuous numbers. More recently, these algorithms have been modified to consider the more frequently occurring integer case [2].

If, for purposes of illustration, we allow parameters as continuous numbers, the designer's search for a "best" system

design can be visualized as a search over a multidimensional surface for its peak value (we shall assume that, unless otherwise
stated, the performance measure value is more desirable as that
value increases). Such a surface arises when we consider a plot
of some performance measure versus a range of values for each
of one or more selected system parameters. Figure 11.2 shows
a contour plot for the widget FMS example system for the performance measure total-parts-produced versus two parameters,
namely speed-of-carts and processing-time-at-machine-3. Note
that the performance measure increases in value as cart speed
is increased, but decreases in value with change in processing
time. Let us call this relationship between the performance measure and the system parameters a performance function (of the
parameters). It seems reasonable to seek the maximum value of
this function in order to get the best set of parameters for the
given performance measure.

One way to search for the maximum of the performance function is, given any arbitrary starting values of the parameters,

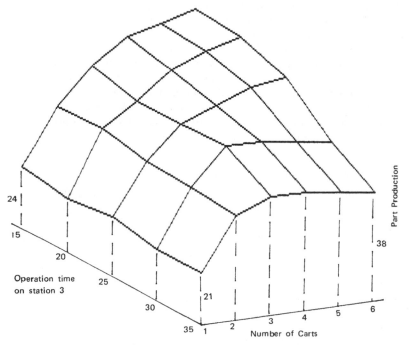

Fig. 11.2 Plot of performance measure for widget example.

to assess the "sensitivity" of the function with respect to each of the parameters. Then, if the function is indeed sensitive to changes in the parameters, modify the parameter values in the direction that promises to move the performance measure nearer to its peak value. The classical way to determine this sensitivity is to obtain a "derivative" of the performance function with respect to the parameters. Since our function is defined via a simulation model and not an analytical function, we shall deal with finite difference, or approximate-derivatives (AD). An AD could be obtained for a performance function and two parameters by making three simulation runs, one at the starting point and then one run for an incremented value of each parameter. Let PSS stand for the performance measure value at the starting point, and let P1S and P2S stand for that value at the incremented values of parameters 1 and 2, respectively. Then, for parameter 1, AD1 = (P1S - PSS)/D1 and for parameter 2, AD2 = (P2S - PSS)/D2, where Di is the value of the increment in parameter i.

These ADs could be used to point the way toward the (possibly local) performance function peak. That is, if AD1 is positive and AD2 is negative, the designer should try a new design alternative with increased value of parameter 1 and decreased value of parameter 2.

The above remarks need to be tempered for the case where significant uncertainty is associated with the system. The ADs then become possibly unreliable estimators of the directions for performance improvement since the values of PSS, P1S, and P2S will vary depending on, for instance, the random-number seeds. (In fact, the notion of a "stochastic" performance function needs to be defined for this case, as in Ref. 3.) However, as shown later, even this stochastic situation can be accommodated by use of more advanced statistical methodology.

11.7.2 Simulation Optimization

A methodology for automating some aspects of the design process when simulation models are available is simulation optimization. This kind of methodology had its beginnings around 1960. The systems of interest in those cases involved mostly chemical processes described via differential equations [4]. The system parameters were of a continuous nature and a performance function was specified. The objective of the methods was to search over the surface represented by this function for its maximum (or minimum, as appropriate) value.

There are two general approaches to accomplish the search.
One is to obtain AD or (if possible) derivative information so as
to lead the search in the direction of the optimum. The second
approach is essentially derivative-less in that information on how
to proceed toward the performance function optimum is obtained
only indirectly and heuristically. Note that, in either case, if
we are considering uncertainty, the response surface is actually
"fuzzy" because there may be many possible performance measure values for a given set of controllable parameters (owing to
the variety of the uncontrollable factors).

The AD-based methods for simulation optimization have been
developed to a greater extent. Let's first consider the situation
where no uncertainty is present in the system. Several useful
methods have been developed for this case which can be applied
to simulation models. The most notable of these for our purposes
is that due to Rosenbrock [4]. He shows that it is indeed possible to find the peak of a function using AD information to modify the values of parameters either positively or negatively. His
method improves on that basic idea, however, by further modifying the directions of movement toward the peak. That is, his
method in effect creates new parameters which are combinations
of the original ones and continues to search in these new directions. Rosenbrock's method gives results for various test performance functions which are improved over prior methods that
do not generate new directions. Still, we can conclude that,
whether Rosenbrock's method or other methods are used, if the
performance function has around five parameters, then the number of simulation runs needed to achieve the peak value is counted
in the hundreds. For over 10 parameters, that number of runs
may be a thousand or more.

A designer might seek the best set of system parameters for
those situations in which one or more of the three kinds of uncertainty mentioned earlier is present. Simulation optimization
methods have also been developed for this case. One of the
earliest such methods was developed by Smith [5]. His response
surface methodology (RSM) is aimed at systems for which there
is a significant uncertainty component. Thus, the performance
function for such systems has a random nature. Obtaining AD
information for such models has increased difficulty in that multiple simulation runs need to be made in order to increase confidence in the performance data. In general, RSM accomplishes
this by using extensions of the basic statistical regression no-
tions of fitting an approximating linear or quadratic surface to

the performance function at each point of investigation. The resulting approximation gives information as to the direction of the optimum of the response function. For example, if the performance function surface is approximated at the search starting point with a linear (planar for two parameters) surface via linear regression, then that approximating surface can be represented by the equation (for two parameters x_1 and x_2):

$$\beta_0 + \beta_1 \times x_1 + \beta_2 \times x_2$$

The coefficients β_1 and β_2 are associated with the gradients of the approximating surface at the starting point. Thus, if β_1 is positive and β_2 is negative, the performance function peak apparently lies in the direction of increased values of x_1 and decreased values of x_2.

Given the gradient information, the RSM methodology moves to a new point (i.e., new values for x_1 and x_2) and repeats the regression process. These iterations continue until all gradients attain small values, which indicates that a peak in the performance function has been achieved.

A second, more recent, AD-type method for simulation optimization, called SAMOPT, was developed at Purdue [3]. This method is based on ideas of Robbins/Monroe [6] associated with stochastic approximations. SAMOPT provides approximate-derivative information on a stochastic performance function, but does so in a different manner than that of RSM. Unlike RSM, the SAMOPT method is guaranteed to converge to the optimum of the usual test (i.e., unimodal) performance functions. Both SAMOPT and RSM have the additional capability of accounting for constraints on the design process (e.g., the total cost for a system to be designed might be constrained to be below some preset value). A comparison of SAMOPT with RSM is given in Ref. 3.

Table 11.3 shows the performance of SAMOPT for several test functions, both with and without constraints on the search. Aside from performance data, the table illustrates a more important point. Note that the number of simulation runs needed to achieve the optimum performance function value with two parameters is on the order of 50−100. This is not at all an atypical result from optimization experiments involving uncertainty. It indicates that automated design via simulation modeling becomes practical when simulation run times are small relative to the time available to make the design decision. It is to be expected that

Table 11.3 Performance of SAMOPT Simulation Optimization

	Actual value	Optimum by SAMOPT	Runs used by SAMOPT
Surface 1	0.8267	0.7859	69
Surface 2	0.7720	0.7362	65
Surface 3	0.8477	0.7753	57
Surface 4	0.8899	0.8869	62

A normally distributed noise with zero mean and standard deviation of 0.10 was superimposed on the four surfaces (all of which have a maximum value of 1.0):

Surface 1

$$y = (.5 + .5x_1)^4 x_2^4 \exp[2 - (.5 + .5x_1)^4 - x_2^4]$$

Surface 2

$$y = (.3 + .4x_1 + .3x_2)^4 (.8 - .6x_1 + .8x_2)^4 \times$$
$$\exp[2 - (.3 + .4x_1 + .3x_2)^4 - (.8 - .6x_1 + .8x_2)^4]$$

Surface 3

$$y = x_1^2 \exp[1 - x_1^2 - 20.25(x_1 - x_2)^2]$$

Surface 4

$$y = (.3x_1^2 + .7x_2^2)^3 \times$$
$$\exp[1 - .6(x_1 - x_2)^2 - (.3x_1^2 + .7x_2^2)^3]$$

The optimization was subject to the following constraints (constructed so that the global optimum falls outside the feasible region for all surfaces):

$$x_1 > = 1.30$$
$$.50 = <x_2 = <2.0$$
$$x_1 + x_2 = <4.0$$
$$x_1 - x_2 = <2.0$$

the more widespread development and use of parallel computers will greatly diminish this run-time problem.

A nonderivative method for simulation optimization under uncertainty, which can also deal with integer values of parameters, has been developed and employed by Lee and Azadivar [2]. The basic search scheme is one created by Box [7], called Complex. Complex employs a regular pattern of points in the parameter space at each of which the system performance is evaluated. A heuristic is used to remove one point from the pattern on the basis of the resultant performance values. This point is replaced by a presumably better one which moves the pattern closer to the optimum of the response surface. Eventually, at least the vicinity of the optimum, if not the optimum itself, is to be the point of convergence.

Since few simulation procedures contain simulation-optimizing programs, it is usually necessary to interface a separate optimization program with the simulation model of interest. Currently, this interface problem is more a function of the base language in which the simulation procedure is written than of the type of simulation procedure (i.e., network, data-driven, or base). Most simulation-optimizing programs are currently written in FORTRAN, and so they are most compatible with simulation procedures with FORTRAN as the base language. For example, the SAMOPT program interfaces rather easily with SLAM and SIMAN models. Interfaces with other languages such as GPSS would seem to require that the programs be linked at the machine language level. Thus, such interfaces are machine-dependent and limited in their portability of application.

11.7.3 Perturbation Analysis

The gradient methods for simulation optimization rely on approximate derivative or finite difference information for their operation. An interesting approach to obtaining the sensitivity corresponding to finite differences has recently been developed and could potentially require only one simulation run to obtain sensitivity, regardless of the number of parameters (recall that three runs were required for two parameters in the discussion of Section 11.7.1). This methodology is referred to as perturbation analysis (P/A).

The perturbation method for obtaining sensitivity measures of system parameters is based on early work by Ho, Eyler, and Chien [8]. In that effort, transfer line manufacturing systems

were the focus. With that paper plus later work, it was shown
to be possible to use the results from only one simulation run to
calculate the sensitivity of the system performance function to
parameters such as the size of each of the interstation buffers.
This method was then generalized to be applicable to a larger
class of simulated systems.

In particular, the "infinitesimal" P/A method [9] applies to
those systems for which the order of the events occurring in
the simulation does not change with a small change in the parameter whose sensitivity is to be studied. Identical "order of
events" in the simulation implies, for instance, that the order
in which customers are served or parts are processed stays the
same, even after the parameter is changed. So if the processing time at some station in the system is to be assessed for its
effect on the performance function of, say, time spent by parts
in the system (note that we would like to minimize this performance function), then small changes in this service time must not
change the order in which the parts are processed in the simulation. Given this assumption, the simulation modeler needs only
to arrange to collect data associated with the time that parts
spend in the system, such as the number of busy times for each
resource object in the system, as well as the number of parts
processed during those intervals. From this specially collected
data, a sensitivity measure for the system parameter, processing time in this case, can be calculated. This, in turn, gives
the designer at least directional information in which to change
the system parameter in order to improve the design. For example, it might be shown that a slightly faster processing time
at a given station significantly reduces the time that parts spend
in the FMS.

In more detail, consider the simulation of a manufacturing sys-
term that contains several stations. We can represent parts
moving from one station to another (and associated processing
time at each station) via a Gantt chart, as in Fig. 11.3a. The
dashed arrows in the figure refer to a move (in zero time) of a
part from one station to another. If we increase the processing
time of station 1 by an increment β, as depicted in Fig. 11.3b,
then each service time, $S1(1)$, $S1(2)$, and so forth, will be increased by β [note that $S1(2)$, for example, refers to the service time of the second part processed at station 1]. Let us take
as our performance function the total time that a part spends in
the system. To obtain the sensitivity of the performance with respect to station 1's processing time, we must, in our simulation,

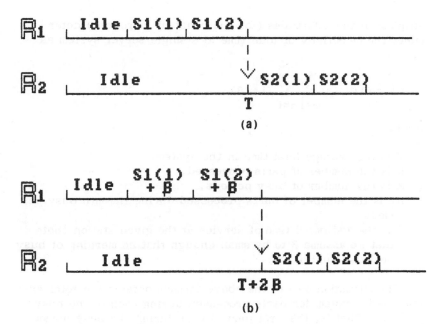

Fig. 11.3 Gantt chart for part movement in P/A.

add data collection for counting the number of busy periods at station 1 (a busy period extends from the time a station ends being idle until the next time it becomes idle again). Within each busy period, count the number of parts processed. The rationale for this data collection is as follows: Any part that arrives at the station of interest spends at least an extra increment of time in the system, namely β. Furthermore, any part that waits for service here must also wait the extra increment of processing time for those parts in front of it in the queue. The result of this propagation of perturbations is that, while the first part to arrive (i.e., to begin the busy period) spends β more time units in the system, the second part which waited for the first part to complete processing starts β time units later and so spends 2β more time units in the system.

A formula may be derived for the propagation of these perturbations over the busy periods. This derivation is done in Ho, Cao, and Cassandras [10] for transfer lines and is given in Suri and Zazanis [11] for M/G/1 single-server queues. The

simplest of these formulas (for the constant increment) deter-
mines the sensitivity of total time in a single-server system as

$$dT = (1/N) \times \sum_{m=1}^{M} \sum_{i=1}^{n_m} (i \times \beta)$$

where

  T is the average total time in the system,
  N is the number of parts processed,
  M is the number of busy periods,
  n_m is the number of parts processed during the mth busy pe-
    riod,
  β is the additional time of service at the given station (note
    that we assume β to be small enough that no merging of busy
    periods would occur)

  The summation on i in the above formula obtains the total extra time in system for parts processed during each of the busy
periods. That is, the first part (i = 1) during the busy period
waits β more time units, while the second part (i = 2) waits 2β
more time units, and so forth. The second summation obtains
the total incremented system time for all parts served over all
busy periods. The 1/N multiplier, of course, produces the averaged value of the time in system over all N customers. Note that,
to calculate dT, it is only necessary to collect data from the simulation on number of parts processed during each busy period as
well as on the number of busy periods. Furthermore, with a relatively small investment in more data collection, sensitivity with
regard to other station processing times could also be obtained
with one simulation run.
  To illustrate the use of P/A, imagine two (admittedly peculiar)
single-server systems. In one, parts happen to be processed at
the given station of interest in busy periods of two parts each,
and for the second system in busy periods of four parts each.
In either case, consider the simulated time during which 24 parts
are processed at the station. In the former case, n_m = 2 for all
busy periods, while in the latter case, n_m = 4 for all busy periods. Then for the two's case,

$$dT = 1/24 \times [12 \times (\beta + 2\beta)] = \frac{3\beta}{2}$$

while for the four's case,

$$dT = 1/24 \times [6 \times (\beta + 2\beta + 3\beta + 4\beta)] = \frac{5\beta}{2}$$

So the system in which parts arrive to the given station in four's is more sensitive re time in the system to changes in the station's processing time. A designer, given this information along with other knowledge such as arrival patterns at the station of interest, may decide to reduce the processing time at that station.

Extensions of the above formula that hold in the more general case of a multiple-station system, where service times, arrival rates, and so forth, are uncertain and so specified by statistical distributions, are given in Suri and Zazanis [11]. Indeed, for certain systems, it can be analytically proven that P/A gives unbiased estimates of the true gradients from a single simulation run.

The present status of P/A is still that of a tool in the development stages. For example, if routing of parts in an FMS depends on the length of queues in the system, examples can be shown in which P/A gives erroneous results. However, although the P/A algorithms are derived by assuming very small (i.e., infinitesimal) parameter changes with no change in event sequence, their sensitivity estimates turn out to be valid for finite, realistic parameter changes. P/A analysis has in fact been applied to FMS systems with capacitated queues on an experimental basis [12]. Though such applications are ad hoc, they do provide useful results when the appropriate assumptions are met.

11.8 GRAPHICAL OUTPUT FROM SIMULATION

Performance analysis for a simulated system employs the simulation output data as its basic source of information. If the system model contains elements of uncertainty, the output data should be treated in terms of statistical design of experiments. This implies collection of averaged data over possibly many intervals of simulated time. So for assessment of performance with significant uncertainty present, it may be misleading to observe the system behavior over a very short time frame. And yet observation over a short time frame is a likely manner of use of graphical output from simulation (one may also lump the "physical simulators" into this category as well).

Fig. 11.4 SIMAN's CINEMA graphical output for an FMS simulation.

    An increasing number of simulation packages provide a graphical display of system behavior as the simulation progresses (e.g., SIMAN's CINEMA, TESS, GPSSPC). These displays provide varying degrees of resolution regarding depiction of the modeled system and its entities. For some packages, the user specifies which displayed symbols correspond to the simulated entities. For example, a graphical output (originally in color) from a SIMAN model of an FMS simulation is shown in Fig. 11.4. Usually, the speed of movement of portrayed entities can be varied by the user, so that they may be viewed in a more comprehensible manner. Other packages provide only a plot of some system variable as the simulation progresses. In either case, the output so obtained could be misleading if the model contains elements of uncertainty. What, then, is the value of these graphical outputs?

Graphical output from the simulation as it progresses can be useful for several purposes. One reason, a sort of "bottom-line" reason, is that the behavior of the simulated system can be shown in more concrete terms to a nontechnical person (e.g., a supervisor or manager). The intention here is usually to gain confidence on the part of that person that the simulation model is indeed a credible counterpart of the real-world system. If credibility is established, then recommendations derived from analysis of the simulation model may stand a better chance of being accepted and implemented.

The other major impetus for graphical output has more to do with the modeler. In particular, the process of debugging a model can be significantly enhanced via some depiction of the modeled system behavior as simulation proceeds. Nongraphical means for doing the debugging include a "trace" of simulated behavior. A trace shows every significant action taken by the simulation program during a specified time period. However, traces are notorious for generating large amounts of data that must be sifted through to find the cause of undesired model behavior. With "on-line" graphical output and a simulation interrupt facility, the modeler can step through simulated time and keep track of pertinent variables as well as model entities. Should a variable take on a questionable value, or should a model object take on a dubious state value, the modeler can stop the simulation and note the current situation. At least this process can establish the rough time at which a problem occurs, and simulation reruns can be made to investigate the problem more thoroughly.

11.9 FUTURE USE OF SIMULATION IN DESIGN OF FMS

Simulation will continue to be a useful and even indispensable design and analysis tool for FMS systems. This is particularly true for the latter stages of design and for the operational phase of the FMS.

The advent of more efficient and generally applicable optimizing programs to be associated with simulation procedures would enhance the ability of the designer to select good system configurations. In particular, the extension of such programs to handle those aspects of FMS configuration that are discrete in nature would represent the most valuable contribution.

Short of the "automated" design iteration situation, there is promise that in the foreseeable future, the combination of artificially intelligent expert systems with simulation modeling will ease the designer's task. This is discussed more fully in Chapter 14.

REFERENCES

1. A. Law and D. Kelton, *Simulation Modeling and Analysis*, McGraw-Hill (1982).

2. Y. Lee, Optimization of Discrete Variable Stochastic Systems by Computer Simulation, Ph.D. Thesis, Univ. of Illinois at Chicago (1986).

3. F. Azadivar and J. Talavage, "Optimization of Stochastic Simulation Models," *Math. and Comp. in Simulation*, 22: 231 (1980).

4. H. Rosenbrock, "An automatic method for finding the greatest or least value of a function," *The Computer Journal*, 3: 175 (1960).

5. D. Smith, "Automatic optimum-seeking program for digital simulation," *Simulation*, 27: 27 (1976).

6. H. Robbins and S. Monro, "Stochastic approximation method," *Annals of Math. Stat.*, 22: 400 (1951).

7. M. J. Box, A New Method for Constrained Optimization and a Comparison With Other Methods, *Computer Journal*, 8: 42 (1965).

8. Y. Ho, M. Eyler, and T. Chien, "A gradient technique for general buffer storage design in a serial production line," *Int. J. Prod. Res.*, 17: 557 (1979).

9. R. Suri, "Infinitesimal Perturbation Analysis of Discrete Event Dynamic Systems: A General Theory," Proc. 22nd IEEE Conf. Dec. and Control (1983).

10. Y. Ho, X. Cao, and C. Cassandras, "Infinitesimal and Finite Perturbation Analysis for Queueing Networks," *Automatica*, 19: 439 (1983).

11. R. Suri and M. Zazanis, "Perturbation Analysis Gives Strongly Consistent Estimates for the M/G/1 Queue," submitted to Management Science, 1984.

12. R. Suri and J. Dille, A Technique for On-line Sensitivity Analysis of Flexible Manufacturing Systems, *Annals of Operations Research*, 3: 381 (1985).

12

Prescriptive Tools for FMS Design

12.1 INTRODUCTION

If a designer of a product or system approaches the design task
with a completely open mind, the number of possibilities for the
entire process are enormous — perhaps impossibly so. More often, the designer uses past experience to guide the choices that
are concomitant with the design process. For example, past experience may suggest a minimum thickness for a product flange,
or it may imply the selection of a milling machine as the preferred
way to make a slot. These experiential biases may be taken as
"givens" in the application of modeling tools to product and system design.

The use of "given" information facilitates the construction of
models that aid the designer. Even for the use of the n-o-q models discussed in prior chapters, some minimal amount of information must be provided by the designer. Thus, the designer must
make choices with regard to at least a few aspects of the thing
to be designed in order to apply modeling to some benefit. Obviously, making a few such choices does not fix the design completely, since many other aspects of the design may not be limited

by those choices. For example, choosing the service time for a
given part type's operation in an n-o-q model still allows considerable leeway with regard to the selection of a machine and a tool
for performing that operation.

The use of "prescriptive" models to aid the designer implies
that some information with respect to the design is given, while
other "controllable" aspects are only specified to lie within a certain set of possible values. Prescriptive models would then be
applied to select the best value of the controllable aspect. Previously we discussed use of n-o-q models and simulation models
in a prescriptive context where the designer employed those models in an iterative manner. In this chapter, we are interested in
models that make the selection of the controllable value without
intervention by the designer.

The set of models of interest here is referred to as mathematical programming (MP) models. In essence, these models are composed of algebraic expressions and equations. The expressions
are the usual ones seen in algebra as constructed from sums and
products of algebraic variables. The equations are of two types,
equalities and inequalities.

Equations in an MP model represent "constraints" that are assumed to hold in the design situation. For example, the acquisition cost of a system may be specified by the designer as having
an upper limit of one million dollars. If we are given the average cost of each machine as c_n and the average cost of each tool
as t_n, then the following inequality would be a designer-specified
constraint on the system design:

$$N \times c_n + T \times t_n \leqslant 1,000,000$$

where N is the number of machines in the system, and T is the
number of tools in the system.

The variables N and T are controllable and might be specified (via other constraints) to lie within certain sets; e.g., N
could be limited to be no greater than 10. The designer could
then use the MP model to calculate the values of N and T that
best meet a designer-specified objective. That latter objective
must be specified as an algebraic expression which is usually
stated in terms of the controllable variables, in this case N and
T. So, if the designer has an objective of creating as small a
system as possible, then the objective may be stated as the min-
imization of the expression

$$N + T$$

In other words, the designer can be interpreted via this model to be aiming at the system that has the smallest number of machines and tools which can still accomplish the production requirements. The production requirements need also be specified in this case as a constraint equation; e.g.,

$$pt1 \times N + pt2 \times N \geqslant 30$$

where $pt1$ is the number of parts per hour of part type 1 that can be produced on a machine, $pt2$ is the corresponding number for part type 2, and 30 parts/hour is the lowest allowable production rate.

The specification of an objective expression and a set of constraint equations constitutes an MP model. It should be apparent that the designer needs to specify new kinds of information in order to use MP models. This new information consists of "targets" for such performance measures as production rates, or of "limits" on available resources such as number of machines.

There are several types of MP models as distinguished by the nature of the variables and algebraic expressions that constitute the model. The most tractable of the MP models are those for which all expressions and equations are linear (i.e., containing no terms that have products or ratios of controllable variables) and in which the variables represent any real number. Once the designer has set up a linear programming (LP) model, he can apply one of several algorithms to solve the model and obtain values for the controllable variables. The most common of these algorithms is referred to as the simplex algorithm [1].

12.2 LINEAR PROGRAMMING

Use of the simplex algorithm to solve LP problems relies on the discovery that the best solution to an LP problem lies on a "vertex of the feasible region." Consider the words "vertex" and "feasible region." If an LP problem has M controllable aspects or variables, then the problem can be viewed as a search in M+1-dimensional space for a best value of the objective expression (i.e., M dimensions for the controllable variables and one more dimension for the value of the objective expression). That search is generally not over the infinite extent of the space, but rather over only a usually finite region called the feasible region. This region is defined by the inequality constraints of the model. That is, each such constraint defines a half-space in which the constraint is true. For example, the constraint

$$N \leqslant 10$$

is satisfied over the entire space below the value of N equal to 10. Given a number of such constraints, their half-spaces may intersect to result in an enclosed region of the M+1-dimensional space. Any point in this region satisfies all the constraints, and so is a feasible (though perhaps not the best) solution.

The feasible region for an LP problem generally has many "corners" or vertices at which the constraints intersect. The best point in the feasible region for any given objective expression in an LP problem must lie at one of these vertices. The simplex algorithm consists of an efficient way of searching these vertices until the best one is found.

From the FMS designer's viewpoint, the ability to find tractable solutions to large MP problems is critical. As shown later in this chapter, real-world FMS design (and operation) problems can be formulated as MP problems. However, it is also the case that such problems are usually of high dimensionality. If MP models are to be useful to the designer, algorithms must be available that are capable of solving large problems.

12.3 MP PROBLEMS IN FMS DESIGN AND CONTROL

Though our main purpose in this book is to consider the design of FMS, the MP tools have been employed sparingly in that context. Rather, they have been used far more extensively to plan and control the production of parts on an operational FMS. We shall consider both aspects of MP use in this section.

In Chapter 7, it was mentioned that the earliest phases of FMS design involve a selection process wherein both the part types to be made on an FMS as well as machines to make those part types must be specified. We can formulate this selection in terms of an MP problem if, as usual, we make some assumptions about the FMS situation.

This statement of an FMS design problem is due to Whitney and Suri [2]. Suppose that we have a list of candidate part types for manufacture on an FMS, and we also have a list of machines that might become part of the FMS configuration. Let the part types have process plans consisting of sequences of "operations" where in fact each such operation is really a stay of some duration on a machine (during which perhaps several tools might be applied to the part). Each part type also presumably has a required annual production volume and is assigned a cost of production, which is the cost of the part type when produced by conventional facilities.

Our formulation as an MP problem will be aggregate in nature; that is, it will consider annual production volume of a part type, but will not assign the actual time during the year when that part type will be made.

The MP problem will be expressed as an objective expression along with some constraint equations. The objective expression in this case is a sum of two terms, the first term being associated with the cost of the selected machines:

$$\text{Cost of selected machines} = \sum_i N_i \times C_i$$

where N_i is the number of machines of type i in the FMS, and C_i is the cost of each machine of that type.

The second term of the objective expression represents the opportunity cost of not selecting a part type to be made on the FMS:

$$\text{Opportunity cost} = \sum_j (1 - X_j) \times R_j \times V_j$$

where j is an index over part types, $X_j = 0$ if the part type is produced conventionally and is equal to 1 if the part type is produced on the FMS, R_j is the annual production volume of part type j, and V_j is the value of part type j when produced in the conventional way.

For a part type j that is not selected for manufacture on the FMS, this second term would have X_j equal to 0. In that case, the total value of that part type would be added to the objective expression. Since the objective expression is expressed in terms of cost, we shall choose the FMS design and part type complement that minimize the expression. That is, we shall try to produce the most valuable part types (measured by conventional costs) at the least cost of equipment.

It is important to note that the objective has variables in it that it will be difficult to interpret as having *non*-integer values. In particular, the N_i and the X_j variables could be interpreted as having fractional values only for cases where some machine is used on a part-time basis in the FMS or some portion of the annual volume of a part type is produced outside the FMS. If we remove these cases from consideration, then although this FMS design MP problem (as stated so far) is linear, it is not directly solvable via LP methods which require that the variables be continuous.

With integer values for N_i and X_j, the design problem is still potentially solvable via integer programming (IP) methods. The formulation of problems for solution by IP methods is similar to that for LP methods, namely, as a set of constraint equations and an objective expression. For most IP methods to work, however, the constraint equations also must be of linear form. Consider the constraints imposed by W&S for this problem: Given maximum floor space, the most machines that can be in the FMS is designated as N; so

$$\sum_i N_i \leq N$$

Two other constraints are associated with the operations for each part type. To formulate these constraints, define a new variable related to X_j as follows:

$$X_{ijk} = 0 \text{ or } 1$$

where the value 1 denotes that operation k of part type j is assigned to machine type i.

The first constraint comes about because we want every operation for a part type to be assigned to exactly one machine type, and so we have an equality constraint for every part type j and for every operation k of that part type:

$$\sum_i X_{ijk} = X_j$$

This constraint says that if part type j is to be produced on the FMS at all (in which case $X_j = 1$), then the sum of the X_{ijk} over all the machine types in the FMS must be 1. Since each X_{ijk} can only be 0 or 1, this implies that for each operation, there is one, and only one, machine type that performs that operation.

The second constraint associated with operations has to do with the total time during a year that machines will be available for processing. That is, the total operation time assigned to a machine type cannot exceed the time that the machine type will be available. If t_{ijk} is the time needed for operation k of part type j on machine type i, and T_i is the time during the year that a typical machine of type i will be available, then,

$$\sum_{j,k} R_j \times t_{ijk} \times X_{ijk} \leqslant N_i \times T_i$$

The left side of each of these constraints sums up the products of the annual volume of each part type j times its operation times on machine type i, if it is to be produced on the FMS.

Finally, there are some tooling constraints for this design problem. These constraints come about because current and prospective FMS incorporate machines that have a local finite storage of tools in a "magazine." Most such magazines are composed of from 30 to 90 "slots," and a tool may occupy one or more such slots. Tooling constraints both for design and for planning/control problems present considerable difficulties.

The difficulties resulting from tool constraints arise at several levels of detail. At the level of minute detail, a given tool might occupy more than one slot, causing some nonlinearities to enter into planning/control problem formulations that we shall see later. At a more aggregate level, and for the design case in particular, a given tool might be shared by different operations on the same machine. Again, this causes a loss of a one-to-one correspondence between slots and tools.

Mindful of these problems, Whitney and Suri settled on the following tool constraint for each operation k:

$$\sum_{i,j} S_{jk} \times X_{ijk} < N_i \times S_i$$

where S_{jk} is the number of slots required for the tools that perform operation k on part type j, and S_i is the total number of slots provided by a machine of type i. This constraint is an

aggregated one, and so might include in the count of slots used
(on the constraint's left-hand side) more slots than would actually
be used. For example, if a tool is used for three different operations on a given machine, its tool slots are counted three times
rather than once. Thus, this MP approach to this design problem
would tend to give a conservative design (i.e., one that will work,
but may be inefficient).

As is often the case in the MP approach, the problem formulation given above is not easily solved for a practical design situation. That is, the W&S formulation is given as an IP problem.
For a real-world design situation with hundreds (even thousands)
of part types to be produced and tens of machine types on which
to make them, the IP problem can take a very long time (and/or
large amounts of computer memory) to solve. In the case where
such problems need to be solved only occasionally, as for long-
range planning, multihour computer run times may only be incon-
venient. But for the design situation where it is desired that the
response time to a designer's request be measured in minutes, solution of the IP problems may simply not be practical.

What does "solution" of the IP problem imply? Essentially, it
means that every possible combination of part types and machine
types has been (at least implicitly) considered before selecting
the best combination. One means of making problems formulated
in terms of IP more tractable is to obtain "approximate" solutions.
In essence, this approximation often consists of considering only
"good" combinations of the controllable variables, but not all combinations, where "goodness" is measured by some heuristic function. The types of such heuristics are many, and detailed discussion of them is beyond the scope of this book. The next two
sections, however, give the general flavor of such approximate
methods to solve MP problems.

12.3.1 Approximate Global Solution Using LP

The development of the IP problem in the previous section for selection of part types to be produced and machines on which to
produce them is characteristic of the contribution that mathematical programming makes to the design of manufacturing systems.
As with other modeling efforts, it motivates the user to be specific about required structure and behavior of the system to be
designed. One unfortunate aspect of MP, however, is that the
problem as finally stated may not easily be solved on today's computers in a reasonable time. Approximate solutions, however,
can often be achieved, as indicated by the following example.

Table 12.1 Work Content for Nine Part Types on Five Machine Types

Part type	Monthly rate	Small MC	Med. MC	Large MC	Med. VTL	Large VTL
				Machine types (work content)		
1	560	0.3	0.4	0.6	0.1	0.2
2	100	2.6	2.9	3.3	0.2	0.4
3	90	2.0	2.5	2.8	0.2	0.5
4	230	0.9	1.2	1.5	0.2	0.4
5	10	3.2	3.6	4.1	1.0	1.3
6	10	1.0	1.5	2.0	5.0	8.0
7	200	3.0	3.25	3.25	0	0
8	100	1.1	1.4	1.7	0.4	0.6
9	10	1.0	1.5	2.0	7.0	12.0
			Unit cost (thousands)			
		150	250	400	500	900
			Cost rate ($/hour)			
		30	35	40	27	30

Take the problem as posed in the previous section and, for purposes of illustration, ignore the detailed tooling aspects associated with operations on parts and tool slots for the tools. Instead, consider only that five types of machines can be used in the proposed system and that the system is to produce parts from a list of nine possible part types (see Table 12.1). Rather than requiring detailed routing and operation time information, assume we need only the work content for each part type associated with each machine type. This work content refers to the number of hours of work per unit workpiece that can be done on each machine type. So, general-purpose machines can perform more of the work content than special-purpose machines.

Having relieved ourselves of the tooling constraints, we are left with a problem of the following form:

minimize

cost of selected machines + opportunity cost =

$$\sum_i N_i \times C_i + \sum_j (1 - X_j) \times R_j \times V_j$$

where N_i is the number of machines of type i in the FMS, C_i is the monthly amortized cost of each machine, j is an index over part types, $X_j = 0$ or 1 if the part type is or is not produced conventionally, R_j is the monthly production volume of part type j, and V_j is the value of a unit of part type j when produced in the conventional way, subject to the constraints

$$\sum_i N_i \leqslant N$$

and

$$\sum_j R_j \times w_{ji} \times X_j \leqslant N_i \times 210$$

where w_{ji} is the work content of part type j on machine type i, and each machine is assumed to be available 210 hours per month, and

$$X_j \leqslant 1 \text{ for all } j$$

The above constraints require that the designed FMS does not exceed N machines in size, and that the work assigned to the selected machines does not exceed their (monthly) capacity.

For cases where the number of part types plus the number of machine types is on the order of 100, the above problem can be solved directly by existing IP methods. However, one study at Draper Labs was concerned with a manufacturing facility that had thousands of part types that could be made on an FMS. This problem size is prohibitive and led to the search for approximate solutions.

One attractive approach to approximate solutions of this sort of problem is via "constraint relaxation." In particular, let us relax the constraints that the number of machines must be integer, and that the part type will be produced either entirely on

the FMS or not at all on the FMS. Thus, the values of N_i and
X_j are now allowed to be (nonnegative) real numbers. The above
problem now becomes easily solvable via LP (this is also the case
even when thousands of part types are under consideration!).
The example data given in Table 12.1 were used to generate a
linear program with the additional stipulation that at most 10 machines were to be employed in the system. The resulting solution obtained was:

Number of machines of each type = $(2.07, 2.74, 3.64, .51, 1.04)$

Part types to be produced on FMS = $(1, 0, .33, 1, 0, 0, 0, 0, 0)$

As expected, the solution is in terms of real numbers which express fractional values for machines and part type production.
This approximate solution can now be subject to further study
to obtain an all-integer solution. Draper Labs researchers [3]
have developed such a procedure which at first essentially rounds
the fractional values for machine types to integers. Then a second step is taken to obtain integer values for the part type results. For our purposes, we can make a first guess by simply
rounding the numbers to integers while checking that no constraints are violated in doing so. The number of machines of
each type could then be calculated as

(Approximate) number of machines of each type = $(2, 3, 4, 0, 1)$

which satisfies our constraint of no more than 10 machines in the
FMS.

If we take our "initial guess" of the part types to be made on
the FMS as part types 1 and 4, we shall again violate no constraints (though we can expect that the system utilization would
be reduced since we are making none of part type 3).

Constraint-relaxation approaches are widely used to solve MP
problems. As mentioned above, there are many possible embel-
lishments to those approaches which render the solutions more
accurate than just the "first guess" obtained here. In fact, the
"rounding up" that was employed to achieve integer solutions is
not only inaccurate, but also not cost-conscious. Rounding a
real number (e.g., 3.64 machines) up to 4 from its next lower
integer value of 3 implies in this case an additional expenditure
of $400,000! A better formulation of this problem might first try

of the machine resources. That is, a calculation for each part
type is made as follows:

$$\text{Relative savings} = \frac{\text{Conventional cost}}{\text{Weighted work contents}}$$

where the "weighted work contents" penalizes the part type that
uses the scarcer resources. That denominator is calculated as
follows:

$$\text{Weighted work contents} = \sum_i (\text{percent of resource utilized}$$

$$\times \text{ work content})$$

where i ranges over the machine types in the system.
   For example, for part type 1, the relative savings would be

$$16800/(650/1890 \times 0.6 + 0/210 \times 0.2) = 81,415$$

This turns out to be the largest value for the remaining part
types, and so part type 1 is selected on this iteration. The next
iteration selects part type 4, after which the LMC resource reach-
es too near capacity to accommodate the next possible part type.
So, the sequential method finally selects part types 7, 1, and 4
for production on the FMS.
   The sequential method implies that the assessment of desirability is repeated over and over again. If an "outer loop" to specify
number of machines is also used as suggested above, then many
combinations of selected part types and machine types will be considered, and so it is anticipated that the final selection will be a
"good" one. All such algorithms are subject to a fundamental limitation, however. That is, the early decisions are made without
knowledge of what the later decisions are going to be, and so
may later prove to be suboptimal. Of course, algorithm devel-
opers are well aware of this fact and have incorporated schemes
that seek to alleviate the potential shortcoming.
   The FMS design situation, then, can employ MP methodology
to aid the designer. Heuristic approaches can provide good approximate solutions to selection problems in the desirable response
time frame of minutes as opposed to waiting possibly hours for an
exact solution.

Table 12.2 Work Content for Nine Part Types on Two Machine Types

Part type	Monthly rate	Machine type	
		Large MC	Large VTL
1	560	0.6	0.2
2	100	3.3	0.4
3	90	2.8	0.5
4	230	1.5	0.4
5	10	4.1	1.3
6	10	2.0	8.0
7	200	3.25	0
8	100	1.7	0.6
9	10	2.0	12.0
		Cost rate ($/hour)	
		40	30

be made iteratively in an "outer loop" which modifies the number of each type of machine on each iteration). The method proceeds by calculating the "relative savings" of each part type not yet selected for production. This criterion is computed from the conventional-production cost (i.e., monthly production rate × work content × cost rate). Initially, we could simply select the part type with the largest conventional cost as one of the types to be produced. In this example, that rule results in the choice of part type 7.

With at least one part type selected, the number of hours available on each machine type may be reduced. In this case, let 210 hours/month be the nominal available time per machine, so that the machining centers initially have 1890 hours available and the corresponding number for the VTL is 210 hours. With the inclusion of part type 7, the LMC available hours is reduced to 1890 - (200 × 3.25) = 1890 - 650 = 1240 hours. The VTL hours are not reduced.

To select from the eight part types remaining, the relative-savings criterion is modified to account for current availability to reassign operations to existing machines that have spare capacity before opting for an additional machine in the system.

12.3.2 Sequential Approach to Prescriptive Design

Another example of an approximate solution to an IP problem is the second tack taken by Draper Labs for the above FMS design problem. This solution method is a sequential one. That is, instead of waiting until all combinations of controllable variables have been considered before selecting the best one (as is, at least implicitly, the case for the classical IP solution methods), the second approach is to make selections of controllable variable values in a sequential manner. So some selections of, for example, part types to be produced on the FMS will be made at the very beginning of the decision process when, in fact, very little information is available to do so. Each such decision is then "local" as opposed to the "global" solution obtained by the classical method.

At the heart of the sequential method is the proposed assignment of a work content from a given part type to a given machine type. This proposed assignment is assessed with regard to its desirability considering its characteristics and those of all the other assignments made so far. The characteristics associated with an assignment include the time it requires a resource to be occupied, the tools it uses, and some economic measure (the economic measure used by the Draper algorithm, called "relative savings," includes simply the saving that would accrue by ceasing to produce the part the conventional way). In general, the desirability criterion should favor the assignments that save relatively large amounts of money while at the same time discouraging assignments that use up scarce resources.

To illustrate the sequential method, consider the data in Table 12.2. Nine part types are described there in terms of their work content on two types of machines, namely large machining centers (LMC) and large VTLs. For simplicity, it is assumed that there is no overlap in work content between the two machine types for any of the part types. The monthly production rate for each part type and the cost rate of conventional production are also given.

This sequential method requires that the machines have already been specified for the FMS. Suppose that a decision has been made that there will be nine machining centers and one VTL. The selection of part types to produce on these machines remains to be done (in fact, the specification of machines for the FMS might

12.4 MP METHODOLOGY FOR FMS OPERA-
TIONAL PLANNING PROBLEMS

Though this book is focused on system design problems, discussion of MP methodology would be incomplete without some consideration of its use for operational planning. Such planning refers to the decisions required to operate an FMS facility that is already in production. These decisions are made to resolve various problems such as selecting the orders that will be filled over the forthcoming time period, loading the machines with tools (for those FMS in which machines have tool magazines), and sequencing of parts into the system. Each of these problems has a combinatorial nature and so may at least be formulated (though perhaps not solved) as an MP problem.

As an important example, consider the second type of problem mentioned above, that is, to assign the tools into the machine magazines based on the part types to be produced. Prior to solving this problem, of course, it would be necessary to have selected the part types to be produced simultaneously on the system, as we did in the previous section. Then, to formulate an MP problem for tool loading, an objective expression and equational constraints must be specified. As shown in Stecke's thesis [4], several types of objective expressions can be written for this problem. For our example, consider the objective of minimizing the number of part movements from one machine to another. This could be a desirable aim in some systems since both part travel time between machines and waiting time (for a busy next machine to become idle) could be saved, thus improving productivity. Thus, there is a definite relationship between tool loading and the control of part flow in the FMS. In fact, an FMS scheduling study of Stecke and Solberg [5] showed that, over many kinds of dispatching policies to control flow of parts in an FMS, the best objective on the average for loading the tools to machine a given set of part types was the minimize-movement objective.

To express the objective, let x_{ij} be the operation assignment variable where $x_{ij} = 1$ if operation i is assigned to machine j and $= 0$ otherwise. Suppose there are B operations to be assigned over M machines. Assume that the index i is assigned such that consecutive operations on a part type have consecutive index numbers. So, if i and i+1 represent consecutive operations, then

$$\left| x_{ij} - x_{i+1,j} \right| = 0$$

if i and i+1 are on the same machine j. Conversely, the absolute magnitude term equals 1 if those consecutive operations are done on different machines.

Consider the following definition of N:

$$N = \sum_{i=1}^{B-1} \sum_{j=1}^{M} | x_{ij} - x_{i+1,j} |$$

Since, for consecutive operations done on different machines, the absolute magnitude term contributes twice to N, once for machine j and once for machine j', N represents twice the number of part movements between machines. Our concern here is to minimize the value of N.

The constraint equations for this problem are as follows:

1. Integrality constraint on the x_{ij}:

$$x_{ij} = 0 \text{ or } 1$$

   for all i and all j

2. Each operation is assigned to one machine:

$$\sum_{j=1}^{M} x_{ij} = 1$$

3. A capacity constraint exists where the number of tool slots required by the operations assigned to a machine do not exceed the number of tool slots available in that magazine:

$$\sum_{i=1}^{B} d_i \times x_{ij} \leq t_j, \quad j = 1, \ldots, M$$

   where d_i is the number of slots required in a tool magazine by the tool(s) for operation i, t_j is the tool magazine capacity for machine j, and t_j is an integer greater than zero.

The final tool capacity constraint has been generalized by Stecke since different operations may require some of the same tools. She shows a capacity constraint that accounts for space saving from multiple use of tools by different operations. That extended capacity constraint is too complex to be shown here, but it is important to note that it is nonlinear in the x_{ij} (i.e., it involves terms that are products of the x_{ij} variables).

This MP planning problem is now seen to be a nonlinear integer programming (NLIP) problem in its general formulation. The nonlinearity exists in the objective expression (owing to the absolute magnitude operator), and the extended capacity constraint, if used, would also be nonlinear. Solution of this NLIP problem in its original form would in general not be a trivial matter. Stecke shows various means by which the problem could be simplified in order to expedite a solution. For example, the nonlinear objective expression can be converted to a linear one (this is a desirable conversion since more efficient solution methods exist for linear problems) by the definition of new variables. The price paid for doing so is that further constraint equations need to be added to the problem in order to define those new variables.

Stecke applied the above minimize-movement formulation to a real-world FMS system. The target system was the Sundstrand Omnicontrol DNC line at the Caterpillar Tractor Company in Peoria, Illinois. As described earlier, this FMS consists of nine machines. including four mills and three drills served by two computer-controlled transporters, and has a 16-station load/unload area located midway along the line's length. The parts machined on the system consist of automatic transmission housings. One such housing required 49 operations. The 49 operations were aggregated by Stecke into nine operation sets which could be performed on the mills and drills (ignoring a few operations on a VTL and inspection), and it was these operation sets that were allocated to the machines for solution of the loading problem. A list of the operation sets and their requirements for machine type and tool slots is given in Table 12.3

The objective expression in this MP planning problem was converted to linear form in order to achieve efficient computation. This conversion resulted in the addition of 16 new variables and 32 new constraints. Furthermore, Stecke used the extended (nonlinear) tooling slot constraint, which originally had 48 variables and 25 constraints. Upon linearization by one of several available methods, 113 new 0-1 variables and 218 new constraints were added to the formulation. At this point, the problem had grown to enormous proportions from a computational perspective. Further work was done to reduce and combine constraints, and the final problem form of 122 constraints and 169 0-1 variables

Table 12.3 Operation Set Data for Selected
Operations in the Caterpillar FMS

Operation set number	Machine type	Tool slots required
1	Mill	26
2	Mill/drill	26
3	Mill	28
4	Mill/drill	17
5	Mill/drill	27
6	Mill	49
7	Mill	26
8	Mill	40
9	Mill/drill	13

was reached. Though manageable on a large computer, such as
the CDC 6600, this is such a large problem that it would not be
reasonable to try to solve it on more than a once-per-day basis.

However, as indicated by prior work of Stecke and Solberg
[5], it may not be necessary to rely on use of the IP solution.
They show heuristics for solving the loading problem according
to the minimize-movement objective. One such heuristic is extremely simple and is of a sequential nature. In this algorithm
(illustrated in an example below), the operations are ordered and
then assigned consecutively to the first machine into which they
fit. This procedure is continued until all operations are assigned.
For this heuristic, the computational requirements are considerably reduced from that of the corresponding IP problem. More
important, Stecke points out that the solution to the Caterpillar
loading problem as obtained by heuristic means was identical to
that from the exact solution of the IP problem.

As an example of a heuristic approach to the minimize-movement loading problem, consider the following algorithm which applies to a given set of part types and their associated operations
and tool slot requirements:

1. Group consecutive operations that require the same machine
    type into sets called operation sets. Consecutively number
    the sets. This has already been done for the data in Table
    12.3.

2. Calculate the number of tool slots required for each operation set.
3. Assign each operation consecutively to the first machine of the correct machine type into which it can fit.
4. Continue assigning operations until all have been allocated.

For the Caterpillar system, a major constraint on the tool-loading problem was that the capacity of the tool magazines was for 60 tool slots. Part of the heuristic for loading that system was derived from the nature of the machines in the system, namely that the milling machines could also do drilling, but not vice versa. Thus, any operation set that could be done on either a mill or a drill, and which followed a milling-only operation set, would be assigned to the mill if the tool-capacity constraint could be met. Stecke further specified this heuristic by requiring that the drills be "pooled"; that is, they would be identically tooled. This assumption implied that the tool-loading algorithm considered only one drill.

Proceeding with the heuristic, take operation set 1 in Table 12.3 and assign it to the first milling machine, say Mill1. Then 26 of the tool slots on Mill1 are no longer available. Continuing to operation set 2, note that it is to be assigned to Mill1 by our discussion of the previous paragraph. At this point, 52 of the available tool slots on Mill1 are used. The next operation, 3, requires 28 slots on a mill and so cannot fit on Mill1. It is therefore assigned to Mill2. The remainder of the heuristic is calculated in a similar manner and results in the following assignment of operations (and tools):

Mill1	Mill2	Mill3	Mill4	Drill
1,2	3,4	6	7,8	5,9

12.5 SUMMARY

Mathematical programming has been used to formulate prescriptive approaches to solving both design and operational problems associated with FMS. Though these formulations are not often solved directly, they do lead the way to heuristic methods which can be efficient and effective.

Several difficulties with the MP approach have already been mentioned (including computer execution time and the representation of complex constraints such as those which result from tool overlap). One difficulty not mentioned so far is the consideration of uncertainty. In our discussion of descriptive methods, the notion of uncertainty was explicit (recall the discussion of random-number seeds for simulation, and the probabilistic basis for n-o-q models). Incorporation of uncertainty in MP models has proved to be difficult, however. Thus, a solution to a given operational problem, for example, might be valid only for a brief time period after which the system status may change beyond the model's assumptions. Procedures such as sensitivity analysis of "optimal" solutions aid in making MP solutions more robust, and there is always the alternative of simply solving the problem more often as the system status changes.

REFERENCES

1. G. Dantzig, *Linear Programming and Extensions*, Princeton University Press, Princeton, NJ (1963).

2. C. Whitney and R. Suri, "Decision aids for FMS part and machine selection," Proc. of ORSA/TIMS Conf. on Flexible Manufacturing Systems, Ann Arbor, MI (1984).

3. *The Flexible Manufacturing Handbook*, The Charles Stark Draper Laboratory, Cambridge, MA (1982).

4. K. Stecke, "Production planning problems for flexible manufacturing systems," Ph.D. dissertation, Purdue University (1981).

5. K. Stecke and J. Solberg, "Scheduling of operations in a computerized manufacturing system," NSF grant no. APR 74-15256, report no. 10, Purdue University (1977).

13

Economic Justification of FMS

13.1 INTRODUCTION

An FMS is a substantial capital investment for any company, and it is natural that, as part of the design task, engineers are asked to quantify the improvement that will be obtained by the use of an FMS. A system (or a number of competitive potential systems) is only hypothetical at the design stage, and so determination of the costs and benefits of such systems cannot be done by observation of performance. Rather, resort has to be made to models from which those measures can be calculated. Since some of these measures are economic ones, an inherent part of the design process involves the use of engineering economic modeling.

Engineering economic modeling is a well-established function in the industrial engineering profession. A textbook in the area would display several types of economic models associated with cash flows [1]. Most of these basic models are applicable to manufacturing system justification problems.

Justification of capital goods is directly related to economic considerations, as stated by Fisher [2]:

The *value* of capital [goods] must be computed from the value
of its estimated future net income, not *vice versa*. This statement may at first seem puzzling, for we usually think of causes
and effects running forward not backward in time. It would
seem that income is derived from capital; and in a sense, this
is true. Income [i.e., services] *is* derived from capital *goods*.
But the *value* of the income is not derived from the *value* of
the capital goods. On the contrary, the value of the capital
[goods] is derived from the value of the income.... These
relations are shown in the following scheme in which the arrows represent the order of sequence — (1) from capital goods
to their future services; that is, income; (2) from these services to their value; and (3) from their value back to capital
value:

Capital goods → flow of services (income)
                           ↓
Capital value ← income value

Not until we know how much income an item of capital will probably bring us can we set any valuation on that capital at all.
It is true that the wheat crop depends on the *land which yields
it*. But the *value* of the crop does not depend upon the value
of the land. On the contrary, the *value of the land depends
upon the expected value of its crops*. [Italics supplied.]

Equipment justification usually proceeds by showing economic
benefit resulting from the use of one type of system over another. Benefit is determined in terms of the present value of
estimated future net income. That is, future income from use of
the equipment is projected and may result in the following estimates (adapted from Ref. 1):

End of year	Income ($)
0	0
1	1000
2	1200
3	1400
4	1600
5	1800

If one assumes an effective interest rate of 10%, the present
equivalent of these cash flows is

$$P = [\$1000 \times (0.9091)] + [\$1200 \times (0.8264)] +$$
$$[\$1400 \times (0.7513)] + [\$1600 \times (0.6830)] +$$
$$[\$1800 \times (0.6209) = \$5163.02$$

where the multipliers in parentheses are the present worth factors for the assumed 10% interest rate. Of course, the value of P is NOT the amount *received* now, but merely the present equivalent when the time value of money is 10%/year compounded yearly.

Present value of cash flows is used in such well-established justification methods as "payback analysis" and "rate of return." The latter are discussed in more detail in Section 13.3.

In Chapter 3, 11 benefits were discussed that arise from using FMS rather than conventional batch manufacturing techniques. Some of these can be directly costed in financial terms. This is particularly true of the costs of floor space and of system hardware where, as a result of high equipment utilization in FMS, less floor space and fewer machines are needed. Fewer machines and the operating procedures of FMS mean reduced manning, and this also gives cost savings. In the main part of this chapter, the use of economic models with this type of directly quantifiable cost saving will be considered. The more difficult to quantify savings are considered subsequently.

13.1.1 NC and CNC: Directly Quantifiable Benefits

The comparison here is of the use of conventional NC machines (one operator per machine) versus CNC machining centers versus flexible manufacturing cells and systems. Let us consider the salient features of each of these production resources.

NC machines were initially designed and introduced to produce highly complex parts consistently. This gave improved quality and a reduction in scrap, but required accurate machines with reliable control systems. For the machines to be controllable in a consistent manner, their whole design required improvement compared with the then existing machines. This led to new designs of machines that were not only controllable to fine limits but were also more powerful. These designs exploited advances in carbide cutting tool technology and cut faster in terms of both the quantity of material being cut and the speed of cutting.

As the quality of machine designs improved further, designers found it possible to widen the specification of machines so that drilling, milling, and boring could all be carried out on one machine. These became known as machining centers. Different

operations require different tools, and to speed the changeover
of tools, automatic tool changers were introduced on machining
centers. From about 1968 onward, computers gradually took over
from hard-wired controllers. Almost all modern machine tools are
now controlled by a computer in some way, giving rise to the term
computerized NC, or CNC.

CNC brought some substantial improvements to the operation
of NC machines. It enabled programs to be edited quickly at the
machine, eased the maintenance and repair of controllers in a number of ways, and opened up the possibility of linking machines together through their computers to form parts of larger systems.
But one of the purported benefits of CNC that has not been realized is related to machine manning. In theory, CNC machines
should be capable of operating by themselves once the start but-
ton has been pressed. However, CNC machines are generally still
not able to monitor their own performance. Management still pre-
fers to have a skilled operator mind each machine as a kind of pro-
tection of their considerable investment.

The notion of multimanning, or having one person take respon-
sibility for more than one machine (including the possibility of
running machines unattended), forms the basis for the several
approaches to economic modeling discussed in this chapter. Multimanning manifests its benefits in several ways. The most obvious is a reduction in direct labor (over conventional methods of
production) to achieve some production rate for given products.
Another benefit is to retain the same labor force, but allow the
machines to operate by themselves during normally nonproductive
times (e.g., lunch hours, break times, operator cleanup times,
third shifts). Both these views of multimanning will be seen as
we consider the kinds of economic models reported in the literature.

13.2 FLEXIBLE MANUFACTURING MODULE

As a first step in introducing FMS, a company may decide to purchase a small cell in order to obtain experience of multimanned
manufacture before committing to a full-scale FMS. A minimal
cell will consist of one machine with an associated pallet-handling
system, workpiece delivery device (e.g., railcar), and pallet
storage system. This minimal cell has been referred to as a flexible manufacturing module (FMM) by Primrose and Leonard (P&L)
[3].

Since an FMM has only one machine, it is appropriate to justify it with respect to CNC machines as the alternative. In order to do so, P&L summarized several studies on machine utilization to arrive at the following table of time percentages for machining functions:

Machining	58.3
Reset tools	4.0
Breakdowns	6.7
Misc.	2.9
	———
Subtotal	71.9 (both CNC and FMM)
Setting	9.6
Inspection	3.6
Wait work/tools	6.1
Misc.	8.8 (e.g., operator absence)
	———
Subtotal	28.1 (CNC only)

The two subsets of functions are separated to show that, since the FMM could work automatically and continuously, the second subset would not apply to the FMM. Thus the machining time available for FMM would be

$$\frac{100}{71.9} \times \text{CNC machining hours}$$

Otherwise, the characteristics of CNC and FMM are assumed to be similar, for example, the same utilization when running, no difference in load/unload times, same time consumed for breakdowns or to reset tools.

The only other assumed difference between CNC and FMM was in their operating strategies. The FMM was assumed to operate during meal (and other) breaks as well as to finish off current work at the end of the shift. This extra running time amounted to an increase of about 9% over the CNC hours. Combining this factor with the factor above for increased FMM time results in the FMM assumed to be doing the work of approximately 1.52 CNCs.

P&L display several scenarios for comparison of CNC and FMM. For example, in one case a company has a requirement for a large number of CNC machines and a significant number would be purchased each year. The basic equipment selection choice might be between an FMM and two CNCs (according to the calculation in the previous paragraph, it would take more than one CNC to do the work of one FMM). Their calculations show the FMM in that scenario having a slight capital cost disadvantage (around 13%). This disadvantage is to be compensated by a reduction in equivalent labor costs. That is, if it is allowed that two men will be required for the FMM, then the equivalent number of men for the CNCs would be 2 × 1.52, or 3.04 men. This savings of 1.04 men/year more than compensates for the initial cost disadvantage, and in fact results in an internal rate of return of 34.2%. With 15% being taken as the after-tax cost of capital, the purchase of FMM rather than CNC is justified in this P&L scenario.

13.3 MODELS TO AID ECONOMIC JUSTIFICATION

An FMS may be taken to consist of three or more machines served by a pallet-handling system, a pallet delivery system, and pallet storage. The whole system will be managed and controlled by a computer. Cost justification methods for FMS employ the fundamental economic concepts of rate of return and present worth, but calculation of the cost savings on which to apply these concepts is a considerably more complex task. Arriving at simple ratios of, say, CNC machining hours to FMS machining hours as was done for FMM is generally not feasible, mainly owing to the interactions among FMS components and among workpieces moving about the FMS. Instead, the economic models are augmented by simulation and n-o-q models which predict the performance of the FMS.

Leimkuhler [4] has carried out a quite comprehensive economic analysis of FMS versus conventional manufacture. His study encompassed several categories of factors including

1. System design factors, such as product characteristics, processing requirements, equipment and labor requirements. system configuration
2. System operating strategies, including product mix determination, number of shifts worked
3. Economic environmental factors, including tax credit provisions, specifications for useful life of equipment, payback requirements

Leimkuhler's approach is to develop relationships such as cost versus output for a given part or product mix. By comparing the relationships for different types of production systems, he hopes "to identify breakeven points where a particular type of system is economically preferred over other systems for particular products or product-mixes." To demonstrate this approach, he considers a given prismatic part and compares its manufacture by a process plan based on conventional machines to its manufacture via a process plan suited to FMS.

In particular, the conventional system (let us call it CS) is taken to be made up of six types of machines, whereas the FMS is assumed to have various numbers of only one type of multifunction machining center. The FMS is also taken to have an automated transporter system under computer control for moving palletized fixtures among load/unload stations and the machining stations. This type of FMS is said to provide "a sharper contrast to the conventional production line" since all operations can be performed at any of the FMS stations. Regarding manpower, each CS machine is assumed to require one operator, whereas in the FMS, one operator is required at each loading station and for each group of four machining stations.

Production estimates for both CS and FMS were made using an n-o-q model, namely Solberg's CAN-Q model. Starting with a minimal system, successively larger systems were considered by first identifying the bottleneck station and then increasing that station's capacity (i.e., increasing the number of machines) so as to increase production. For every configuration including the minimal one, the number of workpieces in process were increased until an output near the maximum was obtained (see Chapter 9, on n-o-q models, for description of maximum output). In this way, a reasonable estimate was obtained for the capital and labor requirements in order to achieve an output capacity that is near the maximum that each system configuration can deliver. Some of the results of these production estimates are given in Table 13.1 (for CS) and Table 13.2 (for FMS).

Comparison of Table 13.1 and Table 13.2 shows that FMS requires a larger equipment investment and a smaller work force (less than half); i.e., the CS is relatively "labor intensive" and the FMS is relatively "capital intensive."

Four different approaches to economic comparison between CS and FMS were demonstrated in terms of

1. The years to pay back the additional investment in the more capital intensive system (assuming that it is also less labor intensive)

Table 13.1 Resource Requirements and Performance of CS for a Part Type

Number of machines by type						Output	Equip.	No.
Type 1	Type 2	Type 3	Type 4	Type 5	Type 6	rate (parts/hr)	cost (×$1000)	of men
1	1	1	1	1	1	2.557	1800	6
1	1	1	2	1	1	3.319	2300	7
1	1	1	2	2	1	2.721	2500	8
1	1	1	2	2	2	4.052	2650	9
2	1	1	2	2	2	5.146	2900	10
2	1	1	3	2	2	6.254	3400	11
2	2	1	3	2	2	6.752	3900	12
2	2	1	3	3	2	7.306	4100	13
2	2	1	3	3	3	7.604	4250	14
2	2	1	4	3	3	8.093	4750	15
3	2	1	4	3	3	9.281	5100	16
3	2	2	4	3	3	10.009	5300	17
3	2	2	5	3	3	10.377	5800	18
3	2	2	5	4	3	11.372	6000	19
3	2	2	5	4	4	11.966	6150	20

Output rate assumes enough workpieces in the system to maintain near-maximum output.

2. The internal rate of return on the added investment before taxes are taken into account
3. The rate of return after taxes
4. The average cost per part produced

Payback analysis measures the number of years needed to "pay back" an initial investment of amount I, which is supposed to generate an annual earnings (or savings in our case) of amount J, where that number of years is given by the ratio I/J. Payback ratios of 1–4 years are often required for automation projects [5]. Our interest is in comparing two systems with capital costs of K1 and K2, where K2 > K1, and with labor requirements L1 and L2, where L1 > L2. Then

$$I/J = (K2 - K1)/(L1 - L2) \times H \times 2000 \times S \qquad (1)$$

Table 13.2 Resource Requirements and Performance of FMS for a Part Type

Number of stations			Output rate (parts/hr)	Equip. cost (×$1000)	No. of men
Load	Transport	Machine			
1	1	2	1.391	1500	2
1	1	3	2.079	2000	2
1	1	4	2.721	2500	2
2	1	5	3.477	3000	3
2	1	6	4.148	3500	4
2	1	7	4.692	4000	4
2	1	8	4.942	4500	4
2	2	8	5.551	5000	4
2	2	9	6.161	5500	5
2	2	10	6.959	6000	6
3	2	11	7.653	6500	6
3	2	12	8.336	7000	6
3	2	13	8.965	7500	7
3	2	14	9.424	8000	7
4	2	15	9.915	8500	8

Output rate assumes enough workpieces in the system to maintain near-maximum output.

where H denotes the hourly labor cost and S is the number of 2000-hour-shifts worked per year. The denominator of Eq. (1) represents the annual savings in labor costs.

Tables 13.1 and 13.2 provide the raw data for the K_i and L_i values needed to calculate the I/J payback ratio obtained by comparing the CS with the FMS. A graph of equipment cost or number of men versus output rate for either table would show a linear relationship. Therefore, it is reasonable to represent each of those relationships via a regression line. This results in four linear equations, where Q is used to represent output rate.

For CS:

Equipment cost (in thousands) = 457 × Q + 808

No. of men = 1.43 × Q + 2.82

For FMS:

  Equipment cost (in thousands) = 818 × Q + 288

  No. of men = 0.7 × Q + 0.61

One can then plot a curve of payback ratio for the additional investment in FMS over CS versus output, as given in Fig. 13.1.
That figure assumes a $20/hour labor rate. The curves in Fig.
13.1 are plots of the following equation (for payback ratio):

$$\frac{[(818 \times Q) + 288 - (457 \times Q) + 808] \times 1000}{20 \times 2000 \times S \times [(1.43 \times Q) + 2.82 - (0.7 \times Q) + 0.61]}$$

They show that the payback ratio increases with output rate and
is quite sensitive to the number of shifts worked.

    Internal rate of return provides another cost measure asso-
ciated with a system or project. It is also a discounted cash flow
method, but it seeks to avoid the choice of some arbitrary interest rate in determining the time value of money. Rather, it attempts to find the interest rate that will cause the cash flows associated with the project (in our case, initial investment followed
by annual savings) to have a present worth of zero. Therefore,
at this interest rate, the present worth of cash inflows will equal
the present worth of cash outflows. Since that interest rate is
determined by the cash flows of the project *itself*, this rate of
return is called the *internal* rate of return.

    The payback ratio may be used to calculate the internal rate
of return on investment. This rate of return before taxes is
that interest rate Rb that satisfies the following capital recovery equation:

  $$I/J = (1/Rb) \times [1 - (1 + Rb)^{-Y}]$$

where Y denotes the number of years in the life of the investment.

    Most companies, however, prefer to evaluate equipment on the
basis of after-tax return. If B is the amount of money saved before taxes are accounted for [the denominator of Eq. (1)], and if
the tax rate is, say, 50%, and if straight-line depreciation is used
for a life of Y years, then the after-tax savings A are reduced to

  $$B = 0.5[B - (K1 - K2)/Y]$$

So, $A = B/2 + (K1 - K2)/2Y$.

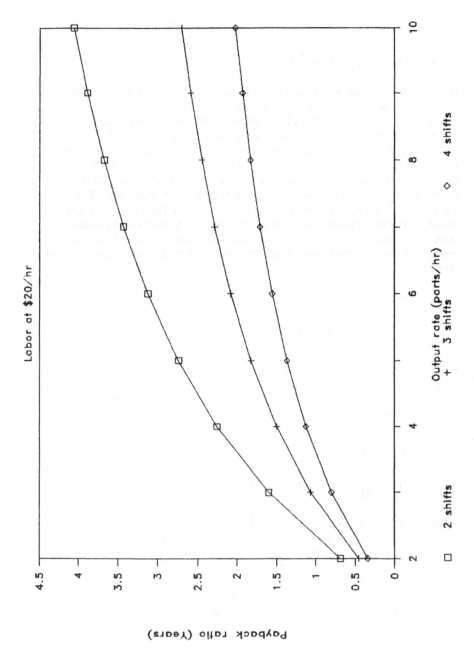

Fig. 13.1 Payback ratio vs. output rate for excess investment in FMS over CS.

Now, A is related to the after-tax return Ra by the equation

$$A = Ra/[1 - (1 + Ra)^{-Y}]$$

By means of these relationships and assumptions, it is possible
to compute after-tax rates of return. Then, in a similar manner to that used for Fig. 13.1, Leimkuhler generates a curve
of after-tax rate of return versus output as in Fig. 13.2.

The last measure to be considered is the average cost of producing a part (or mix of parts). Leimkuhler contends that this
is "the most useful and most informative measure" since "it implies all of the other measures." To obtain this measure, Tables
13.1 and 13.2 were graphed, and it was noted that the relationships between capital cost and labor cost with respect to output
rate were almost linear above certain minimum values of output
rate. These approximations can be shown by the following equations:

$$K_Q = K_O + K^{\curvearrowright} \times (Q - Q_O)$$

$$L_Q = L_O + L^{\curvearrowright} \times (Q - Q_O)$$

where Q is the capacity or maximum output rate for a given configuration.

Then, over the range of these equations, the cost per part
can be estimated by the following:

$$C = (1/Q) \ [(G \times K_O) + (G \times K^{\curvearrowright}) \times (Q - Q_O) + $$

$$(H \times L_O) + (H \times L^{\curvearrowright}) \times (Q - Q_O)]$$

where H is the hourly labor cost and G is the hourly capital cost.

The factor G is estimated by dividing the capital investment by
the payback period in years to get an annual capital recovery cost
rate, and then further dividing that result by $2000 \times S$, where S
is the number of 2000-hour shifts worked by the system each year.

The above equation is used to derive the curves of cost per
part versus output rate for both CS and FMS that are shown in
Fig. 13.3

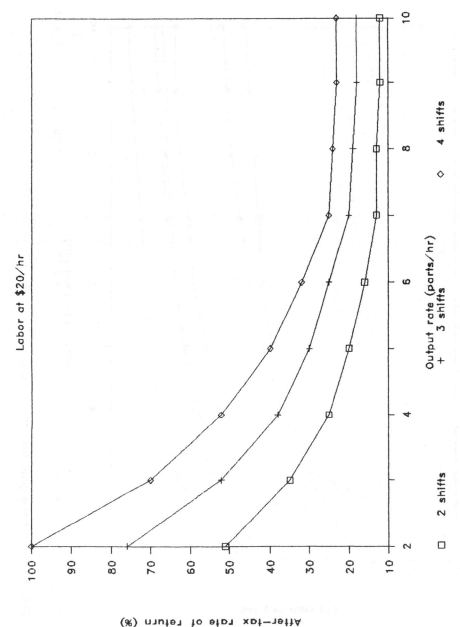

Fig. 13.2 After-tax return vs. output rate for excess investment in FMS over CS.

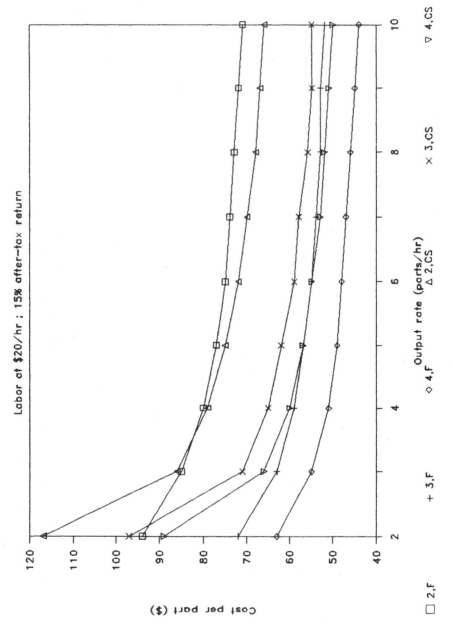

Fig. 13.3 Cost per part for FMS vs. CS.

The advantage of using both the FMS and CS systems on a three- or four-shift basis is clear from Fig. 13.3. Relative to FMS, the CS part cost is generally greater than the FMS cost for three- or four-shift operation, but is not at a disadvantage for production volumes requiring just two shifts.

Leimkuhler states that normally all the "measures of cost are consistent in indicating if one system costs more than another." He summarizes by noting that there is "considerable economic advantage in being able to match a system's production rate to its design capacity by adjusting the product mix. This flexibility to vary the product mix makes it possible to operate a system more intensively and thus get the important economic advantage of higher capital recovery rates."

In another effort toward economic justification, the Charles Stark Draper Laboratory, in their *Flexible Manufacturing System Handbook* [6], indicates the development of a justification package which includes calculation of return on investment for FMS. The estimated production rate for a given system is obtained via computer simulation. In the Draper package, the alternative to building an FMS is taken as subcontracting to outside vendors, rather than using a specified conventional manufacturing line.

13.4 JUSTIFYING FLEXIBILITY

The traditional means of economic justification has been to look for higher production rates and reduced direct-labor costs. With traditional costing methods, direct costs are considered to arise only from the cost of material, machines, and the cost of labor involved in using the machines. But what of the other costs incurred by parts waiting as work-in-progress (WIP) on the shop floor, or the costs of sales lost through long lead times? These and similar costs due to inefficient coordination are not considered at all in most costing systems. Thus, installing an FMS to reduce or eliminate the WIP or long-lead-time costs cannot easily be shown as achieving a saving when no cost is normally calculated for these aspects in the first place. This is discussed more fully in Ref. 7, where Hannam and Kilmartin claim that the real benefits of FMS are in the saving in working capital through reduction of both WIP and product lead times. These real benefits come about, in part, because of short setup times. The short setup times, in turn, allow smaller batch sizes. Primrose and Leonard [3] agree that shortened lead times are among the main benefits of FMS. The result of lower lead time and a reduction in uncertainty of delivery can, from their perspective, increase sales and allow lower stock levels, not just of WIP, but of finished components as well.

Quantification of the benefits of this inherent flexibility property is difficult, but not impossible. Arguably, the benefits should only need to be quantified in percentage improvement terms by a company's industrial engineers. One approach employed the relationship between short setup times (among other things) and small batch sizes in FMS to determine limiting batch sizes below which lower-cost production is "guaranteed" for the FMS over the conventional manufacturing methods [8]. Their tendency to quantify the benefits of the short-term flexibility of FMS in terms of the ability to handle small batch sizes is shown in their summary statement that "the economic advantages of the flexible manufacturing system are brought out all the more by decreasing batch sizes."

Some benefits of FMS are very difficult to quantify. These are hinted at in Leimkuhler's summary statements referring to the advantages accruing from FMS flexibility. The short-term benefits of flexibility are of the type considered by Warnecke and Vettin, while its long-term benefits are of an entirely different character. This long-term approach has been considered by Hutchinson and Holland [9]. The focus of comparison in their work is the benefits accruing to those who invest in an FMS in preference to a transfer line (TL) method for high-volume manufacture. Manufacturers (such as Allis-Chalmers) that have done this have found their FMS investment has maintained its usefulness over far more years than would a TL, thus saving future investment funds for other projects. There are no standard accounting procedures to cost-out such a saving, particularly 5 or 10 years before it may be realized. Hutchinson and Holland have approached this by considering that the introduction of new products over a period of time (instead of consideration of one product for an indefinite time of production, as in the other studies) is the main issue. This concept might be referred to as long-term flexibility. Again computer simulation provides the basis for obtaining the quantification of benefits. New products were created at different times during the simulation and assigned to one of several existing FMS or, if they were fully utilized, to a new FMS. The advantage of FMS in this situation is that it can "incrementally acquire production capacity to simultaneously process many types of parts, and convert production capacity." Apparently, the incremental acquisition of capacity refers to the ability of an FMS to receive additional tooling in order to produce an additional part type. Similarly, the conversion of production capacity seems to imply that existing tooling for a part to be phased out can easily be replaced by tooling for a new part type. This is in contrast to a transfer line which requires full capital expenditure before production can begin, and where production capacity is dedicated to a single part. The results of this justification effort showed an economic advantage for FMS, but more importantly indicated those aspects of the economic and manufacturing environment that significantly affect that advantage.

13.5 ECONOMIC JUSTIFICATION: A DISSENTING VIEW

The total design of an FMS and its subsequent implementation is a very demanding task. There are now enough systems in effective operation for the benefits, both quantifiable and nonquantifiable, to be fully appreciated. The executives of a manufacturing company should have a manufacturing strategy and know whether FMS should be a part of it.

A comprehensive economic analysis and justification of an FMS cannot be carried out quickly and may divert the attention of skilled engineers from important aspects of planning an FMS. In particular, such a justification in *financial* terms may be a waste of time and effort owing to the lack of costing data. Rather, it is suggested that an FMS be justified in terms of its benefits expressed in nonfinancial terms; e.g., percentage reduction in floor space, average utilization level. These figures can also be used as objectives to be achieved during system implementation.

13.6 SUMMARY

Economic modeling is an integral part of the FMS design process. Owing to the interaction of the many components in an FMS, the performance benefits are not conveniently obtained via simple algebraic relationships. Rather, more sophisticated models such

as simulation and n-o-q methods are used to describe system behavior. Given the production rate performance of an FMS, it
has been shown that measures of economic justification by any
of several approaches can be obtained.

Though the quantifications of FMS cost-benefit given in this
chapter show promise, they are not comprehensive in their treatment of the property of flexibility. Since flexibility may be taken
as the *raison d'etre* for FMS, it is important that economic justification be done properly so as not to prejudice an otherwise prom-
ising investment.

REFERENCES

1. L. Bussey, *The Economic Analysis of Industrial Projects,*
   Prentice-Hall, Englewood Cliffs, NJ (1978).

2. I. Fisher, *The Theory of Interest,* Kelley and Millman, New
   York (1954).

3. P. L. Primrose and R. Leonard, "The financial evaluation of
   flexible manufacturing modules," Proc. of 1st International
   Machine Tool Conf., pp. 61–72 (1984).

4. F. Leimkuhler, "Economic analysis of CMS," NSF grant no.
   APR7415256, report 21, Purdue University (1981).

5. T. Prenting and N. Thomopoulis, *Humanism and Technology
   in Assembly Line Systems,* Spartan Books, Rochelle Park,
   NJ (1974).

6. *The Flexible Manufacturing Handbook,* The Charles Stark
   Draper Laboratory, Cambridge, MA (1982).

7. R. Hannam and B. Kilmartin, "Costing, effectiveness and
   financial planning of CNC machines and CNC systems," *CME
   Journal,* June (1982).

8. H. Warnecke and G. Vettin, "Technical investment planning
   of flexible manufacturing systems – The application of practice oriented methods," *Journal of Manufacturing Systems,*
   1: 89 (1982).

9. G. Hutchinson and J. Holland, "The economic value of flexible automation," *Journal of Manufacturing Systems, 1:* 215
   (1982).

14

Artificial Intelligence in the Design of FMS

14.1 INTRODUCTION

The methodology of artificial intelligence (AI) has recently been enjoying a resurgence of general interest after a decline during the 1970s. Though AI methodology has been broadly defined (see Fig. 14.1) to include pattern recognition, machine learning, theorem proving, and modeling of human behavior [1], our interest is in the software developed to support AI research and development. The software for AI has traditionally been represented by the LISP and PROLOG languages.

14.2 LISP

LISP was developed by McCarthy at MIT in the 1950s [2]. It has become the de facto language for use by American researchers in AI. Because of its broad representation capabilities, it can serve as an effective development framework for design software.

LISP is a general-purpose computer language that represents real-world concepts in terms of objects, properties of those objects,

Breadth of Topics in Artificial Intelligence Research

---

* Expert systems

* Knowledge representation

* Programming languages

* Robotics

* Vision

* Natural-language processing

* Machine learning

* Deduction and theorem proving

(For a more complete taxonomy, see the *Applied Artificial Intelligence Reporter*, April 1985 issue)

Fig. 14.1 Topics in artificial intelligence research.

and functions applied to those objects. For example, a LISP object may be taken as a numerically controlled machine that displays such properties as having a motor of a certain horsepower or having a tool magazine of some specified capacity. Collections of objects are represented in LISP by the use of lists (in fact, the name LISP is often taken to stand for LISt Processing language). So, a group of machines in a cell may be represented as

(3_AXIS_NC_MACHINE NC_DRILL INSPECTION_MACHINE)

where the parentheses enclose the objects in the list, and the names of the objects are separated by spaces.

The real power of LISP, however, is not only that its lists denote (ordered) sets of objects, but moreover that LISP allows each list to also stand for a function to be performed! LISP operates on objects by the use of functions. A function can be thought of as an input-output transformation from one object to another. For example, if an object named NC_MACHINE has the property of having a tool magazine, then a LISP function named *getprop* will retrieve the value of the tool magazine capacity via the following command (which is also a list!):

(*getprop* NC_MACHINE capacity tool_magazine)

Since LISP can manipulate lists, including creating new lists, it can in effect program itself. This dual nature for lists gives the programmer tremendous flexibility with regard to representing objects and the functions performed on objects. As a result, LISP turns out to be a very powerful means of knowledge representation. Since our interest in this book is oriented toward models for use in the FMS design process, the notion of knowledge representation is central to our considerations.

14.3 PROLOG

The PROLOG language was developed in 1973 at the University of Marseille in France [3]. It is a computer language associated with predicate logic.

14.3.1 Predicate Logic

Predicate logic [4] is a rigorous approach to describing the world in terms of objects, relationships among objects, and functions performed on objects. In this general sense, it shares the approach to knowledge representation with LISP. An object in logic is again represented by a symbol name. To consider properties of objects, two approaches might be used.

In one approach, a *function* is defined which, when applied to the object, stands for the value of the property that object manifests. For example, we could have a function called *magazine-capacity* applied to the object nc_machine as follows:

magazine_capacity(nc_machine)

This function would presumably map to an integer number representing the number of tool slots for the specified machine.

Logic also allows a way to show the presence or absence of properties. This is through use of a predicate. Predicates may be displayed in the same way as functions, but rather than mapping to any set in general, they map only to two elements: logical "true" or logical "false." Thus, we could consider a predicate to represent the property of a machine having a horizontal spindle, or not, as follows:

horizontal_spindle(nc_machine)

This predicate would map to "true" if the machine named nc_machine indeed had a horizontal spindle, and to "false" otherwise.

The notion of a predicate is, in fact, more general than shown above. It is also used to describe the existence of relationships among objects. If some nc machine, named nc_machine, and a drill in an FMS are on the same AGV path, then the predicate

   on_same_AGV_path(nc_machine,drill)

is true.

The focus of predicate logic is not, however, in the direction of knowledge representation, but rather in deriving *new* knowledge from existing facts. This logical deduction or inferencing capability can be described in terms of sentences in logic that are expressed in terms of predicates. That is, each logical sentence consists of predicates and connectives which tie the predicates together and result in a "true" or "false" sentence. For example, one of the connectives is the operator *and*, which when applied to two predicates

   horizontal_spindle(nc_machine)

   index_table(nc_machine)

in the following way

   horizontal_spindle(nc_machine) *and index_table*(nc_machine)

gives a sentence that is true only if the machine named nc_machine has both a horizontal spindle and an index table.

For our purposes, the logical connective of far more interest is that of implication. The connective operator is called *implies* and may be expressed in sentences of the following form:

   vertical_spindle(nc_machine) *implies does_milling*(nc_machine)

This sentence is often interpreted to say: *If* the machine named nc_machine has a vertical spindle, *then* it can be used to do milling operations.

This kind of sentence, in conjunction with an inference mechanism, can be used to derive new facts. That is, the logical inference mechanism known as *modus ponens* provides that, given a fact such as *vertical_spindle*(nc_machine)

(i.e., the machine named nc_machine has a vertical spindle) and given the above implication, we may conclude that

does_milling(nc_machine)

(i.e., the machine named nc_machine does perform milling operations).

This use of *modus ponens* to derive new facts from existing ones is the basis for the "inference engines" in many of today's AI programs.

Predicate logic has historically been taken as an effective means of describing objects and their relationships via sentences, and as a means of deducing new sentences from an initial set of "true" ones. It has, however, been less useful to represent the control structures for functions that may be performed on objects. It is the controlled selection of functions applied to objects that we often refer to as a computer program. Kowalski [5] and others have now shown that it is possible to display programs in the context of predicate logic. This leads to the notion of a computer language such as PROLOG which implements the concept of logical deduction in its structure.

14.3.2 PROLOG as an Implementation of Logic Programming

The PROLOG language represents objects via symbol names. Relationships among objects, including properties of objects, are shown in the form of "structures," which we can take to be similar to the predicates in predicate logic. So the fact that a machine named nc_machine has a vertical spindle could be shown in PROLOG as

vertical_spindle(nc_machine)

which has obvious similarities to the same fact represented in logic. These facts in PROLOG are connected similarly to the way predicates are connected in logic. In particular, the logical connective *and* is shown in PROLOG by a comma. Furthermore, the connective *implies* is not directly used in PROLOG, but is replaced by its counterpart *implied by*. The *implied by* connector has the special symbol, :- , in PROLOG. So the equivalent of a logical sentence in PROLOG would be

> *does_milling*(Machine) :- *vertical_spindle*(Machine)

This would be read as "any machine whose name is substituted for the variable-name Machine does milling operations *if* it has a vertical spindle."

The reason for reversing the implication connective in PRO-LOG has to do with the type of deduction that PROLOG performs. Its inference mechanism uses implication sentences in their reverse direction (see Fig. 14.2). This is called "backward chaining" or "goal-based deduction." For example, rather than imagining that we are given the fact that a machine has a vertical spindle, we proceed in PROLOG by assuming that we are given a goal of performing a milling operation on some machine. In

---

Implication sentences in logic are often shown with an "arrow" notation as

> *vertical_spindle*(nc_machine) ==> *does_milling*(nc_machine)

Modus ponens may be interpreted as inferring in the direction of the implication arrow. That is, given the "fact" on the left-hand side of the arrow, we may deduce that the predicate on the right-hand side is true. This may be diagrammed as

$$\frac{A \quad A ==> B}{B}$$

Another rule of inference is called *modus tollens*. It can be diagrammed similarly as

$$\frac{\sim B \quad A ==> B}{\sim A}$$

In effect, *modus tollens* allows inference in the opposite direction of the implication arrow. If B is a "goal" to be sought, then *modus tollens* leads to a backward search for conditions that can cause the goal B to be achieved.

Fig. 14.2 Inference mechanism for PROLOG.

order to find a machine on which to perform the operation, we must first search for machines that do milling. But the above PROLOG statement (assumed to be true) states that, if we are to find a machine that does milling, we must at least look for machines that have vertical spindles (of course, other types of machines could also do milling). Then, if it is a true fact that

    vertical_spindle(nc_machine)

the machine named nc_machine becomes one candidate for use in performing the milling operation specified as the goal.

    The capability to represent implication-type sentences and to perform goal-oriented search has resulted in PROLOG being touted as one of the languages in which to write "expert systems."

14.4 EXPERT SYSTEMS

Newell and Simon [6] argued that human problem-solving behavior could to some significant extent be mimicked by a formal system (or computer program) that has at least two characteristics: (1) it uses "production rules" to represent (usually many) simple decisions, and (2) these simple decisions are pursued in the context of a goal-oriented search. Their notion of a production rule is that of an IF-THEN type statement with one of the following forms:

    IF conditions THEN action

    IF premises THEN conclusion

It seems clear that the implication-type sentences in PROLOG could represent the concept of production rules. Of course, their form in PROLOG would be reversed; for example,

    conclusion IF premises

Furthermore, the backward-chaining deduction procedure built into PROLOG could be claimed as an analog to the goal-oriented search that Newell and Simon say characterizes human problem-solving behavior. For these reasons, PROLOG has been employed as a language in which to develop rule-based or so-called expert systems.

Expert systems implement the notion of logical production systems. In these systems, knowledge is represented by a database containing facts and rules of an IF-THEN nature. These facts and rules could be written in PROLOG and would have the form of the examples shown earlier. More recently, a great deal of effort has been made to develop languages that provide for a more English-like syntax for displaying facts and rules in an expert system database. Such languages include the ART [7] and KEE [8] packages. A fact in an FMS design database written in one of the newer languages might be

There are 5 machines in the FMS

while a rule in the database may be of the form

IF there are n machines in the FMS, THEN the number of parts in process is $3 \times n$

The expert system should now be able to deduce that there will be 15 parts circulating in the FMS. This deduction will be a new fact that can be added to the database.

Such an FMS design database might *initially* contain only facts related to part types to be produced and machine types on which to perform operations. The goal of the "design expert system" ideally is to modify the database progressively by deriving new facts in order to end up in some database state containing facts that describe the completed design. It would reach that state by the application of an appropriate sequence of inferences on facts and rules. This implies that a search through a set of feasible alternatives has taken place.

Generally, the chain of deductions that practical expert systems make (i.e., the depth of search) is limited. Even with limited search, many tasks useful to manufacturing can be accomplished.

14.5 EXPERT SYSTEMS IN FMS DESIGN AND CONTROL

Expert system methodology has already seen application to manufacturing systems. The ISIS program [9] has successfully performed scheduling of a Westinghouse manufacturing facility. The approach here was different from a mathematical programming attempt to solve this problem. In ISIS, the constraints associated with the problem were sought out rather than assumed to be ignorable. These constraints were used to guide the search for alternative schedules.

The XCON program [10] (originally known as R1) is used on a daily basis to configure computers for Digital Equipment Corp. XCON is a rule-based program implemented in the language of OPS5 [11]. The OPS5 expert system shell operates on the basis of the *modus ponens* inference mechanism. Since this inferencing proceeds in the direction of the implication arrow effectively contained in the rules, it is referred to as "forward-chaining" inference. Thus XCON is not a goal-oriented expert system according to our previous discussion. Regardless, it is a highly successful application, and its corpus of rules is growing in size as experience with its use accumulates. Its success is demonstrated by the considerable proliferation of other expert systems in use at DEC.

Around the world, there is currently considerable interest in other expert system programs for purposes of process planning, marketing, financial planning, and production control.

More directly associated with FMS is the application of the ART expert shell to accomplish the control of an FMS system at Westinghouse [12]. It is interesting to note that the rationale for use of expert system methods here was not so much related to the usual notion of mimicking human behavior, but was more concerned with a quite different aspect of expert system languages. That is, in computer jargon, expert system languages are often referred to as "declarative" languages. A declarative language is one in which a programmer only declares or defines what it is that is of concern. No actual "commands" to the computer are specified by the programmer. Note that the rules and facts specified in an expert system database are also nonprocedural, and in fact only define what is of concern to the expert system. That is, the PROLOG rule for milling operations given earlier could be taken as the "definition"

> any Machine that performs a milling operation is defined to be (at least one of those) Machine(s) that has a vertical spindle

Any procedures that are accomplished in an expert system are performed by the built-in inferencing mechanism. One of the alleged

advantages of this approach is that modifications to the control
system should be easier to make since only redefinitions or new
definitions need be made, and (ideally) no complex command-flows
need to be considered.

With regard to FMS design, the promise of use of expert systems is alluring. Given the current search capabilities of expert
system methodology as well as the limited speed of (sequential,
rather than parallel) computers, it is not likely that complete
FMS designs will be accomplished automatically in reasonable time.
It does appear possible that expert systems could be used to assist the designer, however [13]. This assistance could be provided in at least three ways. In one, the designer delegates
some of the more clearly defined tasks to the expert system.
This would especially be the case where the design choices are
well established and the criteria for selection are clearly defined.
For example, at Purdue, the COMAND program is being developed to aid the design and modification of automated manufacturing systems [14]. As shown in Fig. 14.3, COMAND utilizes
expert system methods in conjunction with simulation models of
the design situation. It is assumed that, in this design situation, the possible reasons for production delays due to physical
devices such as robots, material-handling carts, conveyors, and
so forth, are completely specified beforehand (for example, an
input buffer may be full, movement speed may be too slow, equipment breakdown might occur). The simulation model is then in-
strumented to detect and display any of these reasons for production delay. Furthermore, the actions to take given any delay are also defined (e.g., increase buffer size, repair equipment). Therefore, the deduction capabilities of the expert system in COMAND are adequate to the task of selecting an action
based on simulation output which results in productivity improvements.

A second way in which an expert system could assist design
is to act as a clerk that checks the design for consideration of
pertinent factors. Since every design is subject to many criteria and constraints, it is easy for a human being to overlook
or forget something. The expert system "clerk," however, would
never suffer from fatigue or boredom and could potentially incorporate many of the rules-of-thumb for design evaluation as es-
poused by the designer or his organization.

Finally, the expert system might be used to perform some of
the tedious calculations now required of the simulation analyst.
This is discussed in more detail in Section 14.6.

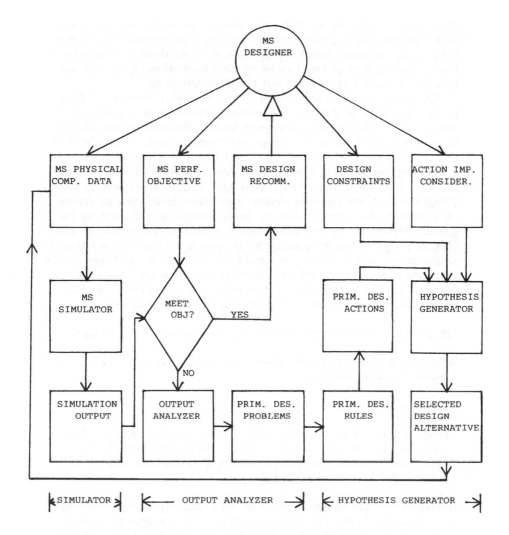

Fig. 14.3 Flow-chart for COMAND combined AI-simulation program.

14.5.1 Explanation Capability for Expert Systems

One expert system development that has been thoroughly documented is the MYCIN effort [15]. The domain of application for this system is that of medical diagnosis. As such, its extension to manufacturing is probably more appropriate for diagnosis problems (e.g., machine breakdown) than for FMS design. Nevertheless, some aspects of the MYCIN system bear directly on the usefulness of expert systems to aid the FMS designer.

One particular feature of MYCIN is of special interest. This is its capability to *explain* the manner in which it arrived at a conclusion. An engineer-designer would presumably have faith in the results of computer programs that perform calculations of the type he was "raised on," e.g., differential equation solutions, matrix algebra manipulations, statistical analysis. It may be more difficult for that designer to accept the output of expert system programs that use rules-of-thumb and other heuristics to arrive at a conclusion. Acceptance of such results would seem to be greatly facilitated by an explanation capability. For example, the MYCIN program can be queried as to how it arrived at its current conclusions, and as to why it needs certain new information to continue its processing. Both kinds of explanation are expressed in terms of the goals that the system is trying to achieve. So we might imagine an expert system to ask the FMS designer,

"Does the machine named nc_machine have a vertical spindle?"

When the designer asks "why do you want to know," the expert system may well respond,

"The fact that nc_machine has a vertical spindle is not in my database."
"That fact is needed to establish that nc_machine can perform milling operations, according to the rule numbered 124."
"That fact is needed to assign a milling operation, which currently is the top-level goal."

Responses of the above type from MYCIN's explanation facility have been available for several years. Such responses can go a long way toward alleviating "Woolsey's syndrome" [16]: "People would rather live with a problem they cannot solve than accept a solution they cannot understand."

14.6 INTEGRATIVE ASPECTS OF AI LANGUAGES

Since the impetus of AI languages is knowledge representation, it should not be surprising that they might be used to integrate several forms of knowledge in one context. This is manifest in current efforts to develop hybrid AI languages. In these contexts, models can be used more easily to represent complex decision making than is convenient now in current simulation packages. In addition, one can represent not only models of real-world systems, but also techniques for "expert" or automated analysis of those models, all in the same framework.

One such language, called KBS [17], was developed at Carnegie-Mellon. This object-oriented language which incorporates a simulation procedure has been used to describe manufacturing systems. Another example of such a hybrid language is the development of SIMYON at Purdue [18]. In addition to an ability to represent rule-based inferencing, SIMYON has a built-in network simulation capability for modeling of systems in the "usual" network way. With its object orientation, it is also possible to conveniently create a new object for any *new* concept that needs modeling. For example, if the system to be modeled includes a scheduling component and none is included among the network symbols of the simulation language, then a new object can be defined that would accomplish that scheduling activity. In fact, the new scheduling object could be a rule-based scheduler that models the behavior of some "expert."

SIMYON not only has the extended modeling capabilities referred to above, but also can incorporate within its framework other tasks that might otherwise require separate programs. For example, the iterative design-by-analysis method discussed earlier requires considerable scrutiny of computer output by the designer. The analysis of performance of a complex system such as an FMS which contains a variety of aspects of uncertainty can be taxing even to the experienced modeler. Yet there appear to be opportunities for some computer assistance in doing this analysis. If a set of rules could be specified for at least some of the activities of the simulation analyst, then the process of performance evaluation could be partly automated or at least computer-assisted.

For example, in a given design or operation situation, the FMS designer/supervisor often has specific objectives that are to be achieved. These objectives in turn imply that certain aspects of the simulation output are more important to the user. So, if he is most concerned about the time it's taking for parts to go through the system, the system could search the output for queues in which the average number or average time exceeds some user-specified threshold. Some portions of this scrutiny can presumably be accomplished automatically by expert system

programs. Such programs could employ rules-of-thumb in order
to accomplish the scrutiny. Since such expert systems can be
implemented as rule-based programs, they can be implemented
in the same SIMYON framework used to implement the simulation
models.

14.7 SUMMARY

The FMS designer, and the FMS operator, can be faced with
choosing from an overwhelming number of alternatives for configuring the system. The modeling tools described in this book
provide a convenient medium for experimentation among those
alternatives. However, the use of modeling tools now requires
expertise in its own right. AI methods offer the promise of incorporating some aspects of the modeling analyst's expertise
into computer programs along with the models of FMS, all with
the objective of improving manufacturing productivity.

REFERENCES

1. P. Winston, *Artificial Intelligence*, Addison-Wesley, Reading,
    MA (1986).

2. J. McCarthy, "Recursive functions of symbolic expressions
    and their computation by machine," *Communic of the ACM*,
    April (1960).

3. A. Colmerauer, H. Kanoui, R. Pasero, and P. Roussel, *Un
    System de Communication Homm-Machine en Francais*, Rap-
    port Group d'Intelligence Artificielle, University d'Aix
    Marseille (1973).

4. W. Hodges, *Logic*, Penguin Books, Middlesex, England
    (1977).

5. R. Kowalski, *Logic for Problem Solving*, North-Holland,
    Amsterdam (1979).

6. A. Newell and H. Simon, *Human Problem-Solving*, Pren-
    tice-Hall, Englewood Cliffs, NJ (1972).

7. *ART (Advanced Reasoning Tool)*, Inference Corp., Los
    Angeles.

8. *KEE (Knowledge Engineering Environment)*, Intellicorp,
    Mountain View, CA.

9. M. Fox and S. Smith, "ISIS — A knowledge-based system for factory scheduling," *Expert Systems*, pp. 25—40, July (1984).

10. J. Bachant and J. McDermott, "R1 revisited: Four years in the trenches," *The AI Magazine*, pp. 21—32, Fall (1984).

11. L. Brownston, *Programming Expert Systems in OPS5*, Addison-Wesley, Reading, MA (1985).

12. "New AI program will schedule in FMS currently being installed," *American Metal Market/Metalworking News*, April (1986).

13. K. Wichmann, "An intelligent simulation environment for the design and operation of FMS," Proc. of 2nd Intl. Conf. on Simulation in Manuf., Chicago, June (1986).

14. J. Talavage and R. Shodhan, "A combined AI-simulation approach for designing manufacturing systems," Proc. of 1987 SAE Intl. Congress, Detroit, February (1987).

15. B. Buchanan and E. Shortliffe, *Rule Based Expert Systems*, Addison-Wesley, Reading, MA (1985).

16. R. Woolsey and H. Swanson, *Operations Research for Immediate Application*, Harper & Row, New York (1975).

17. Y. Reddy and M. Fox, "KBS: An artificial intelligence approach to flexible simulation," CMU-RI-TR-82-1, Robotics Institute, Carnegie-Mellon Univ., February (1982).

18. J. Talavage, S. Ruiz-Mier, and P. Floss, *The SIMYON Manual*, Purdue University ERC report (1986).

Index

T - #0088 - 101024 - C0 - 234/156/20 [22] - CB - 9780824777180 - Gloss Lamination